of

Llyfrgell

Normal

ree short extracts from
ıch is about a different

Continents adrift

Were Europe and North America once part of the same huge continent and have they since drifted apart like icebergs breaking up in the sea? This unlikely sounding idea is now accepted by most geologists today ....

With a pinch of salt

What have tobacco and salt in common? Historians tell us that Cheshire salt used to be heavily taxed, just like tobacco is today. During the wars against Napoleon, salt taxes rose to their peak — and so, it seems, did the activities of the salt smugglers.........

Fig. 1

- How long ago do you think each of these events took place? Match each extract with one of these rough times:
  a few hundred years ago
  a few million years ago
  200 million years ago

Historians are interested in the part of man's past which has been written up in books and documents. Archaeologists go further back than this and look for the bones, tools and buildings of early man. Geologists investigate the history of the earth from its very beginnings, and for most of this time there were no humans around at all (Fig. 2).

Fig. 2
The earth's time-scale

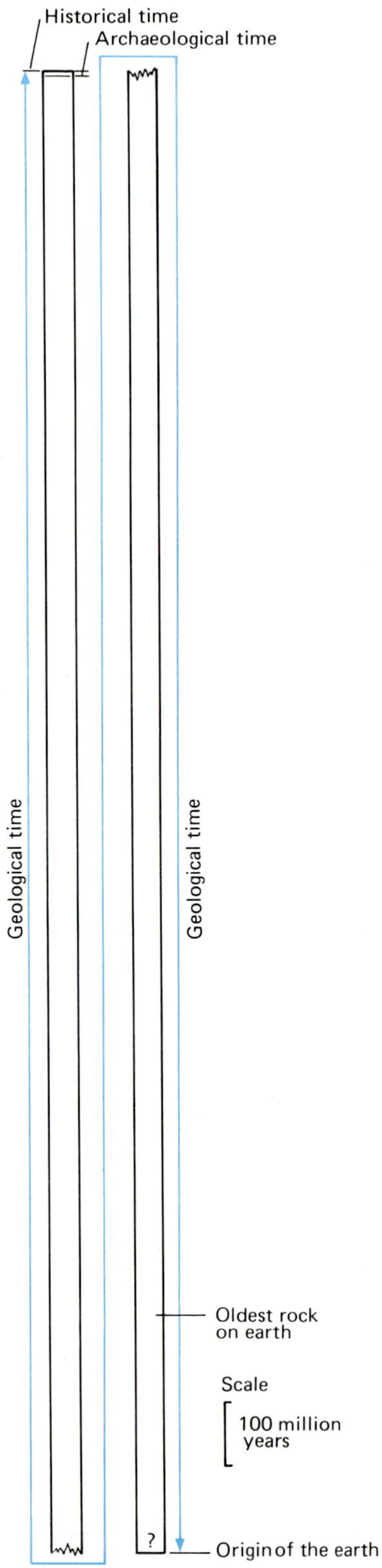

Historical time
Archaeological time
Geological time
Geological time
Oldest rock on earth
Scale
100 million years
?
Origin of the earth

Geologists find their evidence for what happened in the past in stones. There are plenty of these around—on beaches, in the soil in your garden, on mountains and so on. People often talk about stones as if they were really all the same (Fig. 3). But a close look shows that they have different shapes, colours and degrees of toughness.

Fig. 3

Some stones are man-made, like lumps of slag or concrete, and some are natural. All natural stones have one thing in common—they have been broken off rocks at the earth's surface.

The history of the earth is locked up in the rocks beneath our feet. There was a time when people used to believe that all rocks were formed during Noah's flood about 5000 years ago. This simple story is very different from what we now think—that rocks are being formed and broken up all the time, and that these processes have been going on for thousands of millions of years.

What you will find out about the earth's past comes from a study of rocks. But our understanding of what happened long ago is helped by finding out as much as we can about the changes going on today at the earth's surface.

The present is the key that unlocks the history in the rocks. This is why it is important to study the changes brought about by moving air and water, and by earthquakes and volcanoes.

Different kinds of rocks

Rocks in layers

Rocks next to the surface are called outcrops. Often these outcrops are completely covered with a blanket of soil. But Figs. 4 and 5 show two exposures in the banks of a stream in Swaledale, North Yorkshire (Fig. 6). These are places where the soil has been removed to expose the bare rock. You may have seen exposed rock in places like this, and also in crags, quarries and road cuttings.

The rock in Fig. 4 is limestone while in Fig. 5 it is shale. Both rocks have two sets of cracks. One set runs right across the exposures from one side to the other and divides up the rocks into layers (beds).

These cracks are called bedding planes. If you could take all the soil and rock from off the top of one of the bedding planes, it would look something like the top of a table—though not always quite so smooth and flat.

- Use the scales to work out the distances between some of the bedding planes in each photo.

Fig. 4 An exposure of limestone

Fig. 5 An exposure of shale

Fig. 6 Looking downstream at the exposures

Fig. 7
Steeply dipping rocks in a stack on the north Devon coast

- In which photo are the bedding planes (i) more widely spaced, (ii) more narrowly spaced?

Sometimes bedding planes run horizontally across an exposure but often they are on a slope. In Fig. 7 they dip (slope) steeply down to the left.

Fig. 8 A sandstone scarp near Cardiff

Fig. 8 is a photo of a scarp formed by an outcrop of a rock that is more resistant than the rock below it. The dip slope roughly follows the dip in the bedding.

- Is this rock dipping gently or steeply to the left?

The other cracks in Figs. 4 and 5 are called joints. They aren't so continuous and regular as the bedding planes, and they run between the bedding planes roughly at right angles.

- Use the scales to work out the distances between some of the joints in each photo.
- Are the joints evenly spaced in each rock?

How are layered rocks like sandstone, limestone and shale formed? You can begin to find the answer to this question by investigating sandstone.

There is a lot of sand around on the earth's surface today—on beaches and on the sea floor, on the insides of the bends in rivers, and in desert dunes. How can this loose material end up as layers of solid rock?

Activity 1 Looking at a sandstone

Rub a small lump of very easily crumbled sandstone between your fingers so that the grains fall. Look at the separate grains of the rock under a hand lens.

- Are all the grains of the same kind or are there different kinds?
- Do the grains have rounded or sharp edges?

Compare these grains with the grains in a pile of ordinary sand.

Activity 2 Making layers of sand

(a) Add a handful of orange sand to some water in a large jar (Fig. 9). When this has settled, add a handful of sand with a different colour. After this has also settled, add another handful of orange sand.

- What kind of pattern can you see?

(b) Now make a mixture of the two kinds of sand you used in (a). Add a handful of this to some water in another jam jar.

- Can you see the same kind of pattern as in (a)?

Fig. 9

Huge 'handfuls' of material (sediment) are washed down by rivers into seas and lakes. The material comes from the weathering and erosion of rocks.

The action of the weather loosens and 'rots' rocks. These are then dislodged and carried away as stones and boulders by running water, by the wind, or by moving ice. It is these stones and boulders that do most of the wearing down (eroding) of the rocks over which the water, wind or ice moves.

The materials carried in this way are dropped (deposited) when there isn't enough energy to move them any further. For rivers, this can be on the insides of bends or when they reach seas or lakes.

You can tell from Activity 2 that the layer pattern is formed on the bottom of seas and lakes when different kinds of material come into them at different times. The break between the dropping (deposition) of two loads has to be long enough for the upper surface of the first load to harden. The second load then forms on the top of the first without mixing with it.

Fig. 10 shows some graded bedding in a sedimentary rock. Such a rock must have been formed under water because, without water, the material couldn't have been graded like this.

So by studying how modern sediments are deposited under water, we can learn more about the way some rocks were formed millions of years ago. Here is one of the keys that unlocks the history in the rocks.

Activity 3 Making graded layers of sediment

Mix together some gravel, coarse sand and fine sand. Add a handful of this mixture to some water in a large jar. Wait for all the material to settle in a layer (the fine sand may take a few hours to do this).

- How does the grain size of the material change from top to bottom in this layer?
- Can you explain how this pattern has been produced?

Now add another handful of the mixture and wait for it to settle. You now have two layers of graded materials.

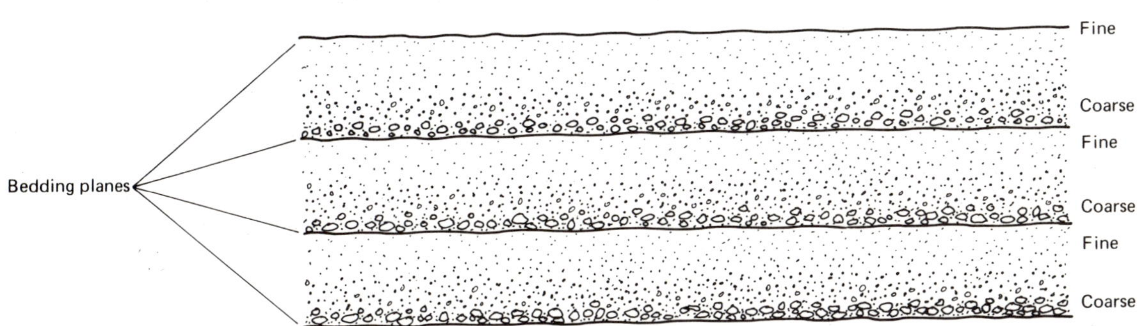

Fig. 10
In each of the three beds, the grains are graded upwards from coarse to fine

The link between rocks and water suggests that there may be something in the Noah's flood idea after all! Though as you will see, not all rocks were formed in this way, and they were certainly not all formed at the same time.

Of course, it is still a long way from horizontal layers of loose sediment under water to dipping layers of solid rock on land. How does loose sand become solid sandstone?

Activity 4 Making 'rock' from sand

(a) Follow the instructions in Fig. 11 using damp sand in a plastic syringe with its nozzle cut off.

- Has squeezing the sand made the grains stick together at all?

(b) Do the experiment again, but use a mixture containing 3 spoonfuls of damp sand with each spoonful of clay powder.

Leave the cylinder of 'rock' to dry out, and then look at it under a hand lens.

- Does the clay help the sand grains to stick together?

(c) Do the experiment again, but use a dough-like mixture of sand and plaster made as shown in Fig. 12.

Leave the cylinder of 'rock' for a few minutes, and then look at it under a hand lens.

- Arrange the 'rocks' formed in (a), (b) and (c) in order of how well the grains stick together. (Put the least easily crumbled one first.)
- Where would the sandstone you broke up in Activity 1 fit into this order?

Fig. 11

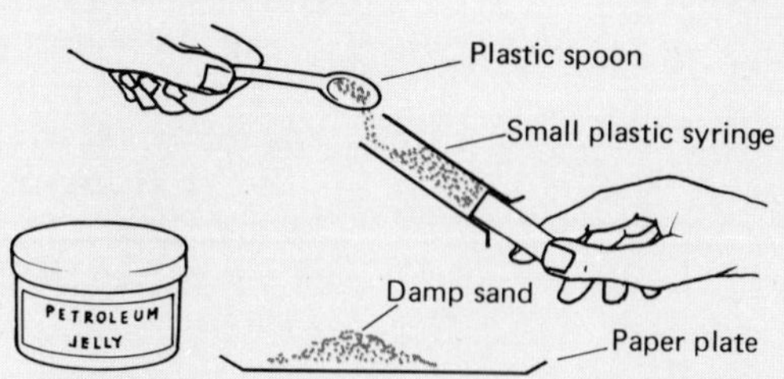

1 Smear the inside of the syringe with petroleum jelly. Then fill the syringe with damp sand.

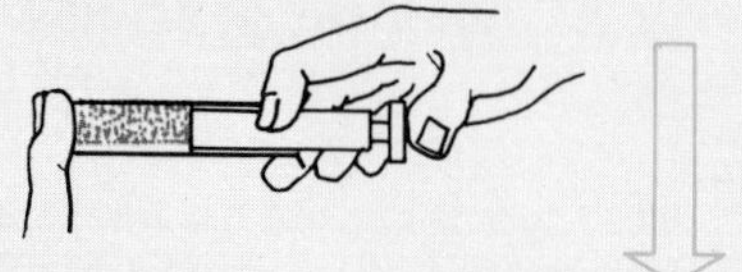

2 Put your thumb over the end of the syringe and press in the plunger as hard as you can

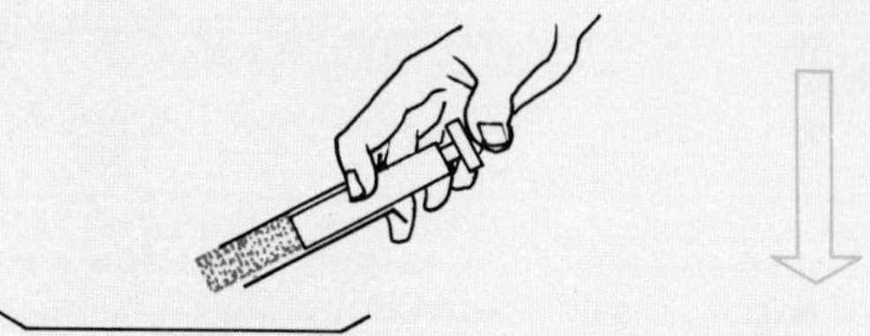

3 Push the cylinder of sand out of the syringe onto a paper plate

Fig. 12

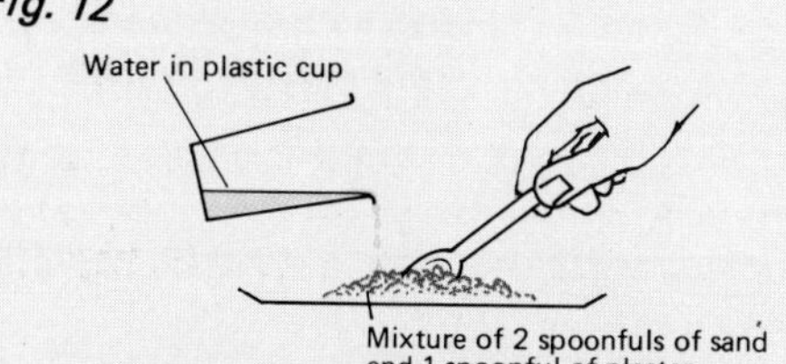

Add enough water to make a dough-like mixture after stirring

Now look at the picture of a sandstone under the microscope (Fig. 13). The rock is made from grains of sand but there is also something between the grains. This is the 'glue' (the 'cement') that holds the grains together. It acts like the clay and the plaster in the 'rocks' made in Activity 4.

Two things happen to the grains of sand in the syringe:

1. they are squeezed together by the syringe plunger
2. they are stuck together by a cement

In nature, the squeezing of the layers of sediment is caused by the huge thicknesses of other layers which are piled on top. Water moves between the grains during this squeezing process. The cement comes from the dissolved substances in this water.

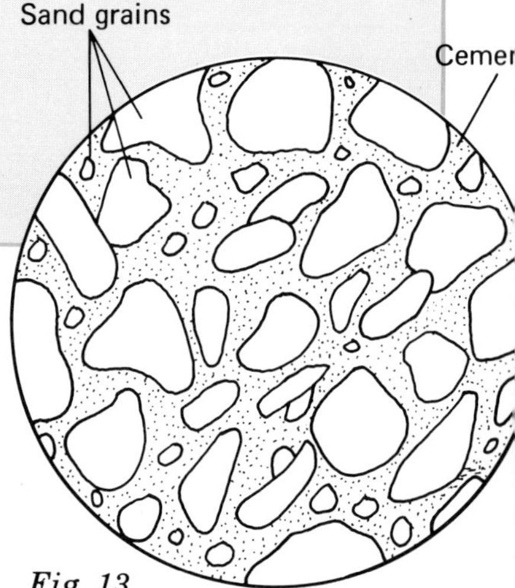

Fig. 13
Sandstone under the microscope (magnified about 30 times)

Another change that takes place during the squeezing of the sediments is that the water trapped between the grains is pushed out. This makes the sediments shrink in size so that cracks may begin to form in them. Some of the joints you can see in layered rocks have been made in this way.

EROSION OF EXISTING ROCKS

Rivers carrying sediment

More layers on top

DEPOSITION OF SEDIMENT IN LAYERS
Seas, Lakes

SQUEEZING, CEMENTING, DRYING OUT (SHRINKING)

Fig. 14
How layered sedimentary rocks are formed

The diagram (Fig. 14) shows what you now know about the formation of layered rocks.

All rocks formed in this way belong to a large set of rocks called sedimentary rocks.

- Can you explain how this name comes about?

Sedimentary rocks can be recognized using the features in Fig. 15.

SEDIMENTARY ROCKS:

+ are almost always found in LAYERS

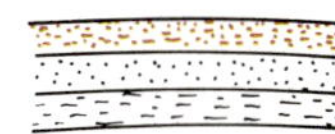

+ are usually made from GRAINS (often stuck together by crystals of a CEMENT)

+ often contain some FOSSIL SHELLS along with the grains

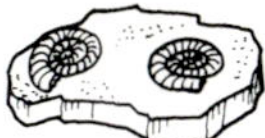

+ are sometimes made completely from FOSSIL SHELLS (often stuck together by crystals of a CEMENT) see Fig. 16

Fig. 15

They often contain whole shells or bits of shells like the ones drawn in Fig. 15 and in the photo (Fig. 16). These are fossils (page 37). Obviously there must have been many living things in the places where these rocks were being formed.

Fig. 16
A shelly limestone

Another kind of layering

Not all sediments are deposited in flat parallel layers. Some have no layering at all, and some have layers that are on a slope right from the start.

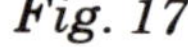
Fig. 17

Fig. 17 shows a way of getting a set of sloping layers on a small scale.

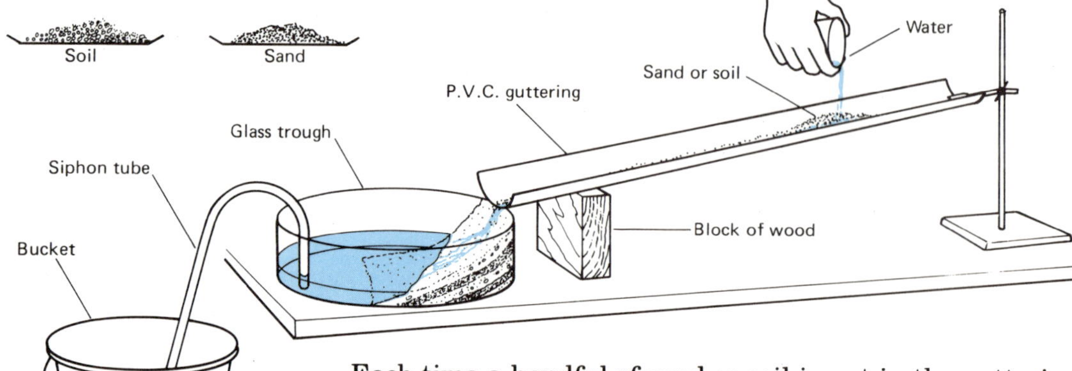

Each time a handful of sand or soil is put in the guttering, water is slowly poured onto it until all of it has been washed into the trough. You can see the layering well if you use handfuls of soil and sand (or different colours of sand) in turn. The layers slope down from the edge of the trough towards the middle.

This kind of layering is called cross-bedding. It can be found in sandstones or gritstones.

Cross-bedding is formed under water when strong currents move the sediment along the bottom of the river or the sea. There is a steep slope at the front of each moving mass of sediment (a ripple or a 'dune'). Grains fall down this slope and make sets of layers (Fig. 18).

Water can deposit both flat layers ('proper' bedding) and sloping layers ('false' bedding or cross-bedding). But the wind can only deposit sloping layers.

Fig. 19 is a photo of the cross-bedding in some ancient sand dunes. The cross-bedding in modern sand dunes helps us, along with other evidence, to unlock the history of these sandstones. They were formed in a desert that covered most of Britain 200–250 million years ago.

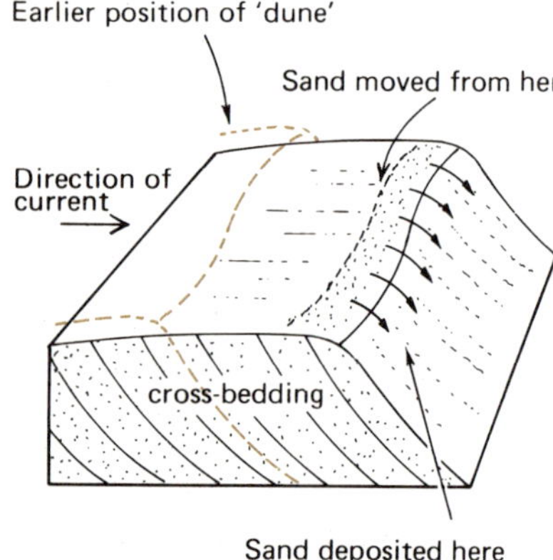

Fig. 18
How cross-bedding is formed

Fig. 19
Cross-bedding in the sandstones at Dawlish, on the south coast of Devon

Fig. 20
A granite sea cliff in Cornwall

'Proper' bedding can also be on a slope (Fig. 7), but it is not deposited like this in the first place. There must be more to tell about sedimentary rocks. The story of how they are raised above sea level and sometimes tilted is taken up on page 25.

Occasionally rocks that weren't formed from sediments can be found in parallel layers. For example, a series of lava flows from a volcano (page 18) can produce the same effect.

Rocks with crystals

The rock in the cliff in Fig. 20 is not a sedimentary rock.

- What does this rock *not* have which sedimentary rocks do have? Look at the cracks in this rock.
- Use the scale to work out the distances between some of the cracks.
- Are these cracks evenly or unevenly spaced?
- Which set of cracks in sedimentary rocks, bedding planes or joints, are they more like?

Fig. 21
Polished granite from Shap Fell, Cumbria

The rock in the photo is called granite. The polished surface of another granite (Fig. 21) shows that the rock is not made from grains. Instead there are interlocking crystals of different shapes, sizes and colours.

- How many kinds of crystals do you think you can see in this granite?

It isn't easy to break up a lump of granite with a hammer because the crystals are tightly locked together, but it can be done with effort.

Activity 5 Sorting out the crystals in granite

Break up a small lump of granite into as fine a 'powder' as you can. First wrap the rock in a cloth, and then hit it with a hammer on a hard surface. Because granite is a tough rock, you may have to work hard to get a 'powder'!

Use a needle to sort out the crystals into three (or four) piles (Fig. 22).

- Make a rough guess of the percentage of each kind of crystal from the size of each pile.

PILE 1

Black shiny flakes

PILE 2

White or pink crystals

(These could be separated into two piles if you find both colours in your granite).

PILE 3

Greyish glassy crystals

Fig. 22 The three (or four) kinds of crystal in granite

Fig. 23
Crystals of halite (rock salt)

Granite is made from three different minerals—quartz, feldspar and mica. Each mineral has crystals with a particular shape and colour.

What is the history locked up in the crystals of granite? Was granite formed under water like most sedimentary rocks?

Fig. 23 shows crystals of a mineral that is found in layered rocks and was definitely formed under water. And many limestones are made from crystals of another mineral (calcite). So it is possible for rocks with crystals to be formed under water.

But there is another way of getting crystals which doesn't use water. This is by melting a solid and then letting it cool down.

Do the crystals in granite come from the evaporation of water containing the dissolved minerals? Or do they come from the cooling of a melt?

- What do you think the absence of any layers in granite means?

Two geologists took up this argument in the late 1700s. Abraham Gottlob Werner, a German, thought that granite was formed under the sea. James Hutton, a Scot, believed that it was formed from a melt and that this melt came from somewhere inside the earth. We now know that Hutton was right.

Activity 6

(a) Follow the instructions in Fig. 25 using salol (phenyl salicylate).

Fig. 25

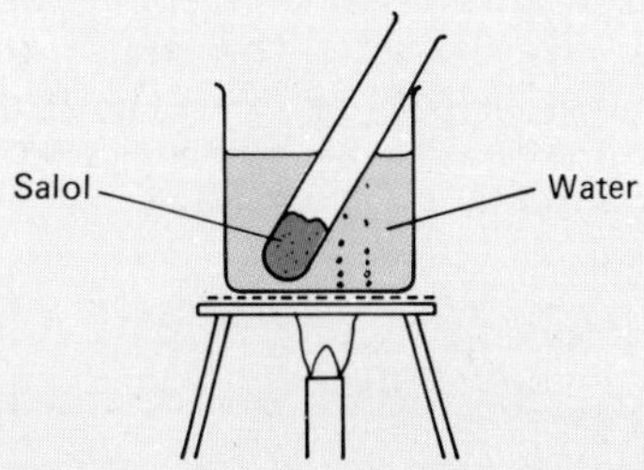

1 Heat the salol in a test-tube in a water bath until it has all melted

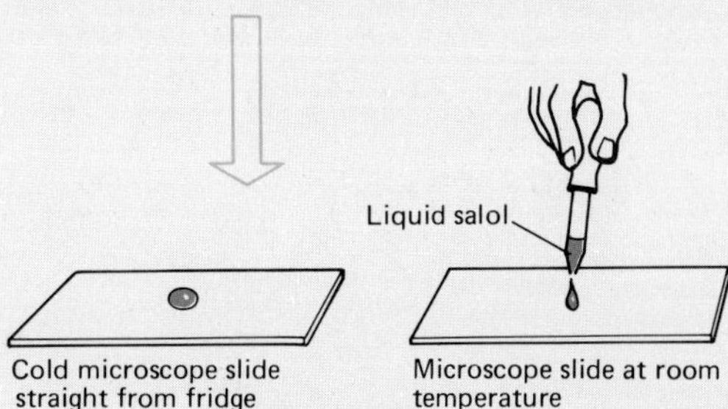

2 Use a dropper to put one drop of liquid onto a cold slide, and one drop onto a slide at room temperature. Watch the crystals growing on each slide through a hand lens

- What difference is there between these two sets of crystals (Fig. 26)?

Fig. 26(a) Salol crystals formed on the cold slide;

Granite is made from large crystals. This means that it must have been formed by slow cooling from a melt.

Everyone knows about one spectacular kind of melt on the earth. This is the kind called lava that flows out of a volcano (Fig. 24).

Fig. 24
A lava flow on Surtsey, Iceland. (See Further study 2)

Making crystals from a melt

- How does this difference come about?

(b) Salol crystals formed on the slide at room temperature

(b) Now find out what happens when a melt is cooled very quickly indeed.

Follow the instructions in Fig. 27 using some powdered roll sulphur.

Look at the solid (called plastic sulphur) that is formed. It isn't made from crystals.

- Why are no crystals formed in this experiment? (Hint: a sulphur melt would form crystals if it were cooled like the salol in (a).)
- Put these solids in the order of how quickly they are formed from a melt. (Start with the one that is formed most quickly.)
  Small salol crystals
  Large salol crystals
  Plastic sulphur

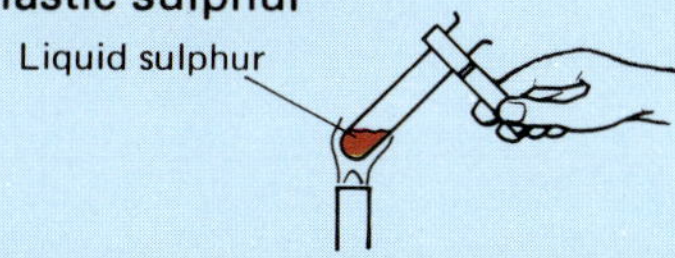

1 Heat some powdered roll sulphur until it turns into a dark red-brown liquid and is just about to start boiling

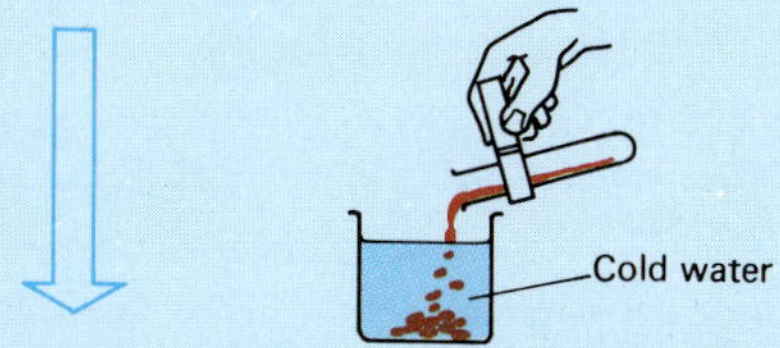

2 Quickly, but carefully, pour the liquid into cold water

Fig. 27

- Use the table in Fig. 28 to work out the rough temperature of the lava in the photo.
- Would you expect this lava to cool quickly or slowly when it meets the air (or the sea if the volcano erupts under the sea)?

You can read more about the speed that a volcano can form rocks on page 15 and in Further study 2: *The birth and death of a volcano.*

- Why can't granite ever be formed by the cooling of a lava from a volcano?

Basalt and obsidian are two kinds of rock that are formed from a volcano. Basalt has very small crystals that can only be seen under a microscope. Obsidian is a glassy rock with no crystals in it at all.

- How quickly must the lava have cooled to form (i) basalt and (ii) obsidian?

Colour of lava	Temperature of lava (°C)
White	More than 1 200
Yellow	900 – 1 200
Red	600 – 900

Fig. 28
The temperature of lava

Fig. 29
Columnar jointing in the Giant's Causeway

Look at the exposure of the basalt lava in the photo (Fig. 29). This is part of the famous Giant's Causeway in Northern Ireland. You will notice the vertical joints. These can be seen in many basalts but they are unusually regular in this exposure.

- How do you think they were formed?

The Ancient Greeks and the Romans had their own way of explaining how lava got to the surface (Fig. 30). They imagined that Vulcan, the god of fire, was working away at melting things down in his underground forge.

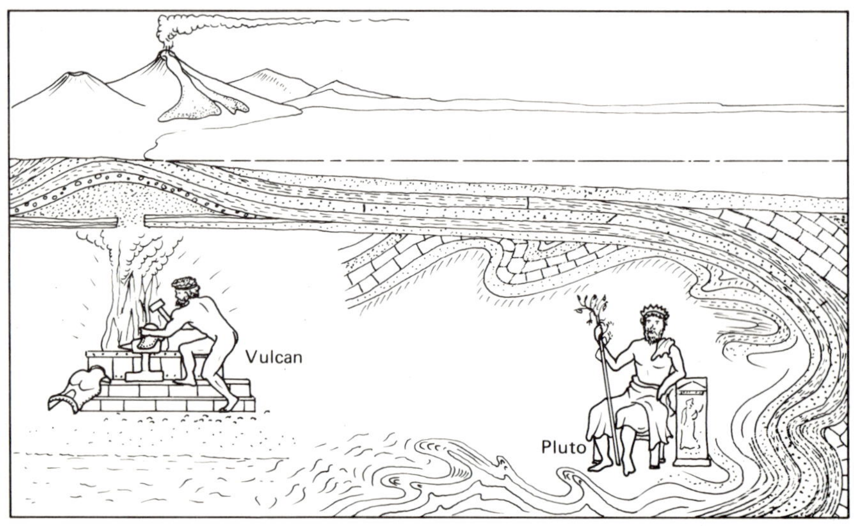

Fig. 30
Early ideas about magma and lava

But there was another, much more sinister, god called Pluto, who was supposed to live in Hell, deep in the bowels of the earth. People used to believe that all dead souls would end up in Pluto's kingdom for ever.

Perhaps this was the fate of the granite. The melt (magma) stayed deep inside the earth all the time it was cooling and forming crystals. Cooling is always slow there because the surroundings are at a much higher temperature than on the surface. It gets fairly hot in Hell!

This leaves us with a problem to solve if we want to finish the story. Granite is formed deep inside the earth and yet can be found in many outcrops at the surface today. Fig. 44 in the next section shows how granite can 'escape' from Pluto's kingdom.

Rocks with crystals that are formed by cooling magma or lava are called igneous rocks. Two kinds of igneous rock are named after the gods Vulcan and Pluto (Fig. 31).

Some intrusive igneous rocks are formed by the more rapid cooling of magma at shallower depths. These are called hypabyssal rocks, and their crystals are of a medium size.

Kinds of igneous rock	Size of crystals	How formed?
VOLCANIC (extrusive)	Small crystals (or none at all – glassy)	From cooling magma that reaches the surface and flows out as lava
PLUTONIC (intrusive)	Large or very large crystals	From cooling magma well below the surface

Fig. 31

Finding the age of a rock

Look back at the exposures of rocks in the stream in Swaledale (Figs. 4 and 5). These are part of a whole set of exposures running all the way up the stream (Fig. 32).

The limestone and the gritstone outcrops are less easily broken up and worn away than the shales. So they stand out from the hillside in a series of small 'steps'.

If you could scrape away all the soil, you would see that these rocks are piled on top of one another in a rock succession. Now that you know all the rocks come from layers of sediment under water, you can tell some of their story.

- Which of the rocks that are named in Fig. 32 must have been laid down (i) first, (ii) last?
- Do the rocks get younger or older as you go up the succession?

Fossils may also help you to unlock the history of a succession of sedimentary rocks. There is more about this on page 43.

It is easy to know with layered rocks that one rock is older than another. But the problem is different with a rock like granite.

Imagine a large mass of granite in a succession of layered rocks (Fig. 33). As the cooling magma pushes its way upwards, the heat 'bakes' the layered rocks around it. The signs of 'baking' around an igneous rock always prove that it must be younger than the layered rocks it has affected.

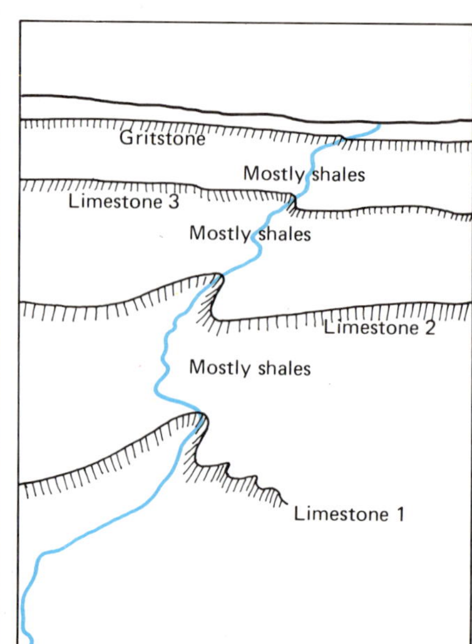

Fig. 32
An overall view of the Swaledale stream shown in Figures 4 and 5

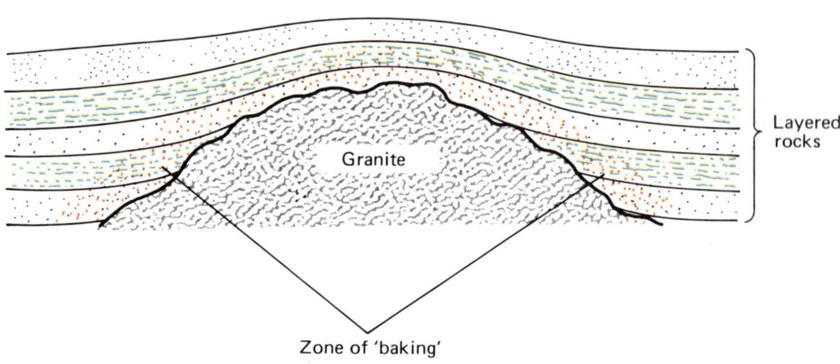

Fig. 33
The zone of 'baking' around a granite

So far, there has been no mention of the ages of any of these rocks in millions of years. The only way of measuring these ages is by using the built-in 'clocks' in some rocks.

Not all rocks have these 'clocks', but igneous rocks like granite and basalt usually do. As soon as the crystals of the granite have been formed, some part of them starts to change to another material. The part of the rock that does this is said to be radioactive. Radioactive minerals always 'tick away' at definite speeds, and so we know how old the rock is (its radiometric age) by finding out how much change has taken place.

Only a few sedimentary rocks have their own 'clocks'. But geologists have still been able to build up a time-scale using the rocks that do have them. So they are able to date a sedimentary rock by looking at the other rocks near to it.

A simple example of this is where layers of a sedimentary rock have a basalt lava sandwiched between them. The lava must have been formed just after the layer below it and just before the layer on top of it. So the radiometric age of the lava is much the same as the age of the sedimentary rock.

The changing earth

A never-ending battle

Not many people are lucky enough to see rocks being born—and also live to tell the tale. But this is exactly what happened to a Mexican farmer, Dionisio Pulido, in 1943.

For a few years before that he had been puzzled by a hole in his cornfield that kept on appearing each time he thought he had managed to fill it in with soil. This problem turned out to be minor compared with having a 400-metre volcano suddenly grow on his doorstep.

Here is a diary of the main events in the first nine years of the Parìcutin volcano (Fig. 34).

All rocks are born to die. There is always a battle going on between the processes forming new rocks and the weathering and erosion which destroy the rocks already made.

Parìcutin hasn't been worn away very much yet. On land, weathering and erosion don't usually work very quickly.

But a more dramatic example of this battle is the volcanic island of Surtsey, off the coast of Iceland, (see Further study 2: *The birth and death of a volcano*). Right from the start in 1963, the pounding of the sea began to wear the island away. This battle was fought between fire and water. Fire won the first round, but the sea steadily took over as the volcano subsided.

Things happened suddenly at Parìcutin and Surtsey. But many processes which change the face of the earth only take place over millions of years. Mountains, lakes and oceans come and go, but we don't notice all that much happening in our lifetime. So people often talk about something being 'as old as the hills' as if to say that the hills have always been there and always will be.

To understand how changes at the earth's surface come about, you have to look at what is happening inside the earth.

Column of smoke rises through ½ metre deep crack

Small explosions of hot fragments from 50 metre cone

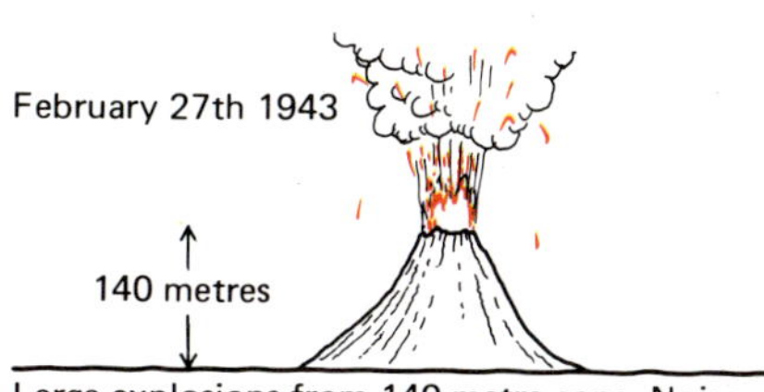

Large explosions from 140 metre cone. Noise heard 350 km away

Mid-1943

250 metres

Lava

Lava breaks through the base of the cone. People have to evacuate the village of Parìcutin 3 km away

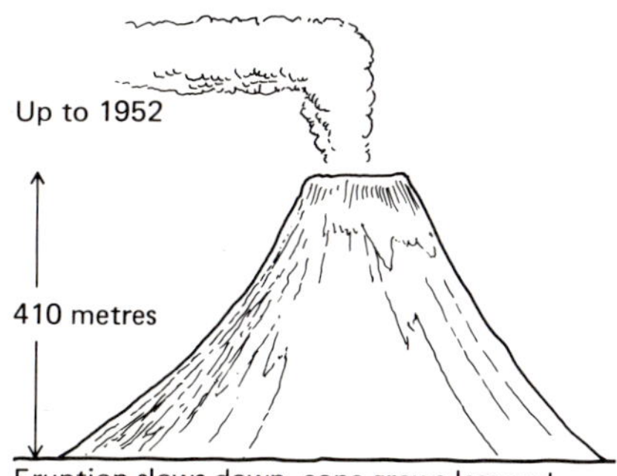

Eruption slows down, cone grows less and less quickly. No more eruptions after 1952

Fig. 34
Some stages in the life of Parìcutin

What is inside the earth?

Man has travelled a long way from the earth into space but he has never journeyed far in the other direction. Though some boreholes go down a few kilometres, these are only pinpricks compared with the 6370 kilometres to the centre of the earth.

Despite this, geologists have still been able to find out a lot about the earth's interior, especially from a study of earthquakes.

Imagine knocking on a wall in order to discover what it is made from. The sound you hear is different for different materials (brick, wood or plasterboard) and for solid walls or hollow walls. Geologists don't go around knocking on the earth's surface. Instead, they pick up and study the shock waves reaching the surface from earthquakes (page 20).

Figs. 35 and 36 show that the crust forms a thin 'skin' of rocks above two other zones, the mantle and the core. The crust under continents is mostly made from granitic rocks and the crust under

Crust
Mantle
Core
3 370 km
5 – 70 km
6 370 km

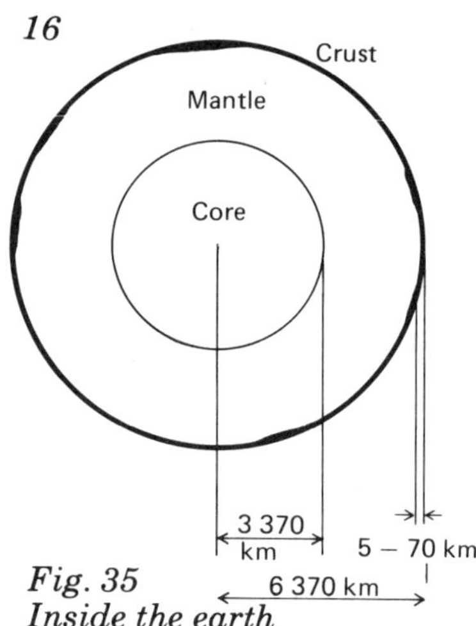

Fig. 35
Inside the earth

Continental crust
Oceanic crust
Ocean
5 km
Ocean
Mantle
Granitic rocks
Basaltic rocks

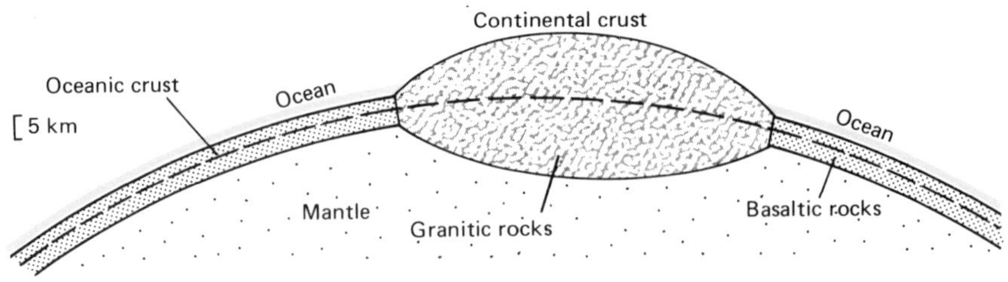

Fig. 36 Continental crust and oceanic crust

oceans is made from basaltic rocks. Granite is only one kind of granitic rock and basalt is only one kind of basaltic rock.

- Use the scale in Fig. 36 to find the thicknesses of the oceanic crust and the continental crust.
- If you were trying to drill down to the mantle, which part of the crust would you choose?

Activity 7 Finding the relative densities of two crustal rocks

Follow the instructions in Fig. 37 using a small lump of granite.

- Use the formula below to work out the relative density of your granite (and so of the rocks in the continental crust).

Relative density=

$$\frac{\text{Weight of rock}}{\text{Weight of an equal volume of water}}$$

Do the experiment again using a small lump of basalt.

- Use the formula to work out the relative density of your basalt (and so of the rocks in the oceanic crust).

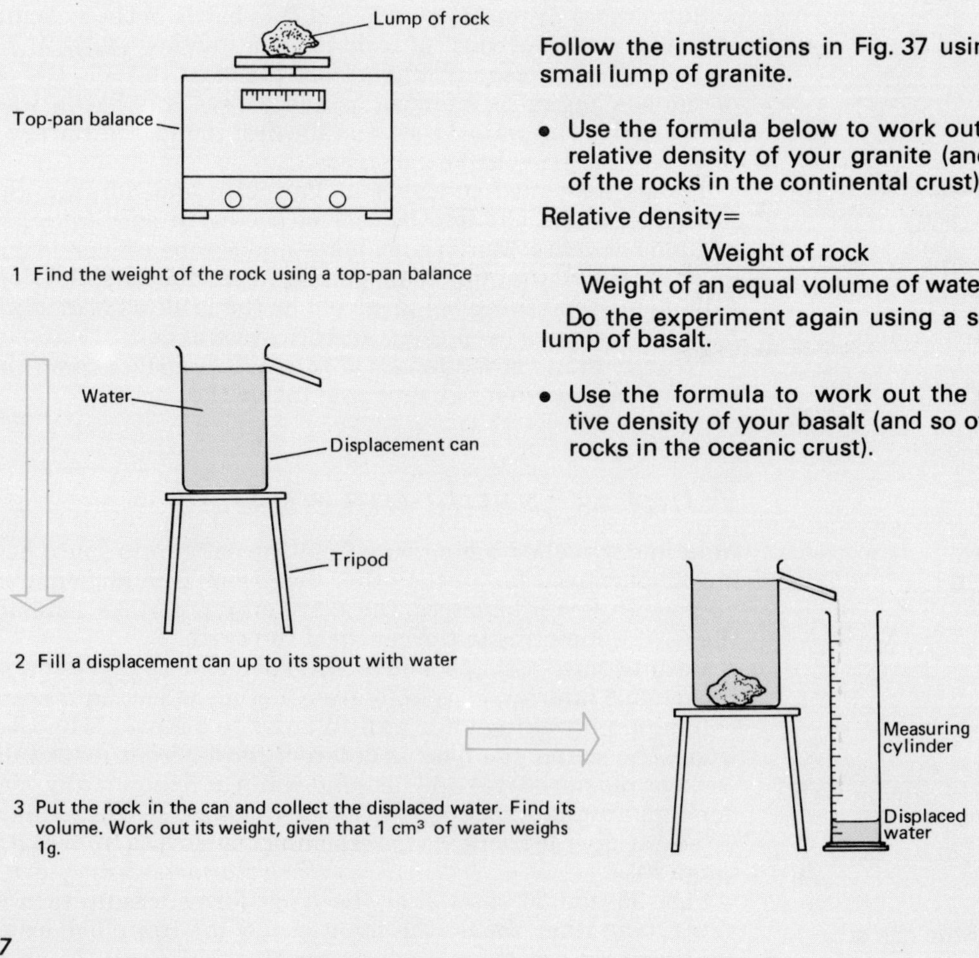

Lump of rock
Top-pan balance
1 Find the weight of the rock using a top-pan balance
Water
Displacement can
Tripod
2 Fill a displacement can up to its spout with water
3 Put the rock in the can and collect the displaced water. Find its volume. Work out its weight, given that 1 cm³ of water weighs 1g.
Measuring cylinder
Displaced water

Fig. 37

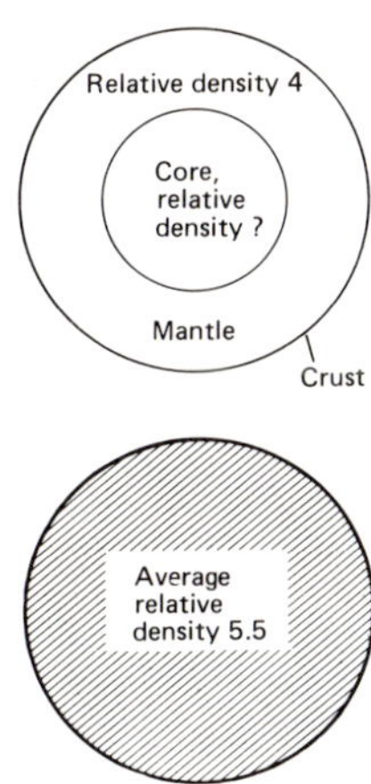

Fig. 38

The average relative density of the whole earth can be measured indirectly—not, of course by the method you used in Activity 7! This turns out to be 5.5, which is much higher than the relative densities of most crustal rocks.

The Americans did set out to drill right through the oceanic crust to the moho (short for Mohorovičić discontinuity or boundary), the boundary between the crust and the mantle. But like many ambitious ideas, the Mohole Project became too expensive, and it was ended in 1966 long before the moho had been reached.

In fact, rocks have since been found in Cyprus which may once have been part of the mantle. They have been pushed up to the surface as Africa crashed into southern Europe (page 24). These rocks can be studied without spending millions of pounds on deep boreholes.

The Cyprus rocks have a relative density of about 4.

- Is the relative density of the earth's core likely to be less than 5.5, about 5.5 or more than 5.5 (Fig. 38)?

The core is probably made from iron and nickel under great pressure. The outer part is a liquid while the inner part is solid.

Any journey into the earth would get hotter and hotter. Look at the temperatures in Fig. 39 for different depths in a borehole.

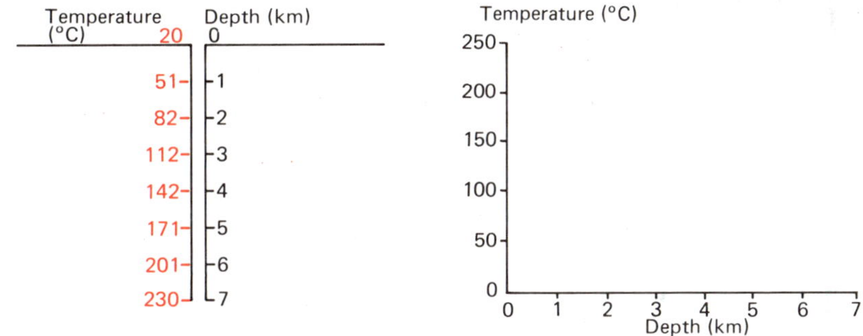

Fig. 39
Borehole temperatures

- Plot your own graph of the temperature versus depth using these figures. Draw a straight line which goes close to or through all the points.
- Use the line to work out how many degrees the temperature rises for each kilometre of depth.

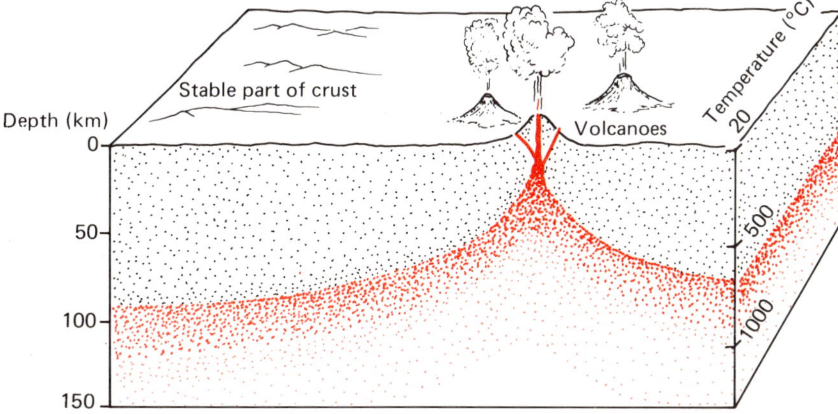

Fig. 40
Temperature zones inside the earth

The temperature doesn't keep on rising as quickly as this all the way to the centre of the earth and the rate varies from place to place. Fig. 40 shows two places on the earth—one where there is an active volcano and the other at a stable part, like Britain, where there are no volcanoes and very few earthquakes.

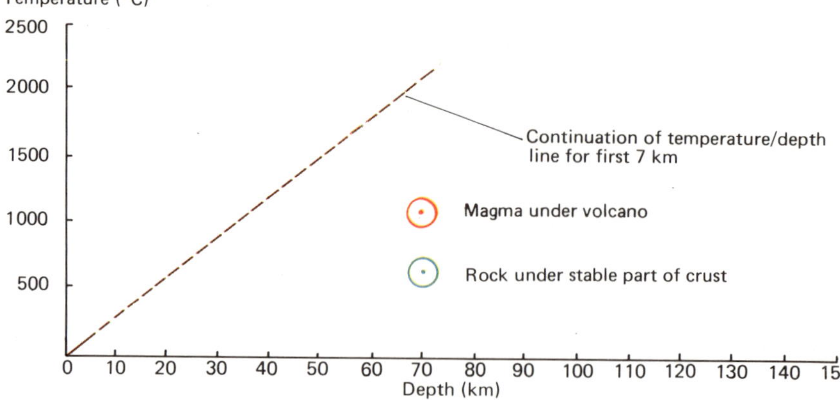

Fig. 41

Look at the new graph of temperature versus depth for greater depths (Fig. 41).

- What temperature are the rocks 70 kilometres beneath a stable part of the earth?
- What temperature are the rocks 70 kilometres beneath active volcanoes?
- Does the temperature seem to go up as quickly after the first few kilometres as it does before this?

There are 'hot spots' between about 70 and 100 kilometres beneath volcanoes. Exactly at what depth the rocks begin to melt depends on the pressure and the temperature.

The melting points of rocks increase with increasing pressure. Since pressure rises steadily inside the earth, deeper rocks are more difficult to melt than rocks close to the surface. On the other hand, temperature rises rapidly at first (Fig. 41) and this more than offsets the effect of pressure at the 'hot spots'.

The magma formed by the melting has a lower density than the surrounding rocks, and so it begins to rise through the crust towards the surface. Not all the magma in a volcano manages to get that far (Fig. 42). Some cools so much that it forms rocks before then, in sills and dykes.

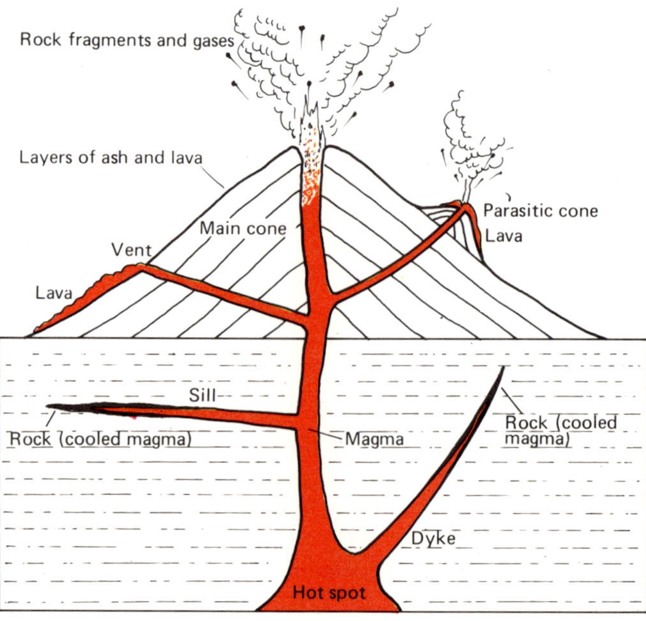

Fig. 42
Some features of a volcano —at the surface and under the ground

There must have been a hot spot under Edinburgh over 300 million years ago. The volcano has long since stopped erupting, but you can piece the story together by looking at the rocks that were formed by it (Fig. 43).

Fig. 43
A view of Edinburgh with the ancient volcano drawn on top of it

- Try to match Fig. 43 with Fig. 42 and so find on Fig. 43 (i) a parasitic cone (a small cone on the flanks of the main cone), and (ii) a sill.

Sometimes the whole of a 'pocket' of magma in a hot spot can't make it to the surface. Instead, it may rise slowly in huge 'blobs' called plutons. These break off and 'digest' bits of solid rock just above them. They cool to form solid granite at a depth of a few kilometres (in Pluto's kingdom, page 13).

A part of the rock cycle

The rock cycle is a way of summarizing the processes that make rocks and mountains and those which destroy them. It is called a cycle because the whole thing happens (very slowly) over and over again.

Fig. 44 shows the part of the rock cycle that you already know about.

The complete cycle is more complicated than this. To finish it, you need to find out how the bits of solid rock on the surface can end up deep inside the earth and perhaps form new magma.

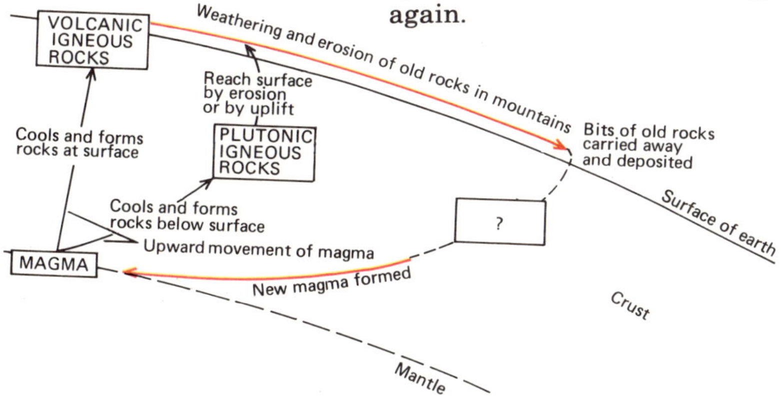

Fig. 44
The part of the rock cycle that concerns igneous rocks

Earthquakes and mountains

Rocks deep inside the earth are under great pressure from those above. A volcano is one effect of a sudden lowering in this pressure. Another effect is that the earth begins to shake. This shaking reaches the surface as an earthquake.

Imagine you are slowly bending a stick until it breaks (Fig. 45). The pressure builds up in the bend until the stick can take no more. The release of this pent-up energy sends sound waves moving through the air and you hear a loud crack.

Just before an earthquake, the rocks inside the earth are under great pressure. The pressure is relieved by the rocks breaking and slipping past one another. The pent-up energy is released and shock waves travel through the earth in all directions.

The table in Fig. 46 is a way of showing how intense an earthquake is. The worst affected area (the epicentre) is always just above where the rocks are breaking.

Fig. 45

Number on Scale	Effect of earthquake
1	Only detected by instruments
2	Felt by people at rest
3	Vibration like passing light lorry. Hanging objects swing
4	Felt by people walking. Windows rattle. Vibration like passing heavy lorry
5	Felt by most people. Sleepers woken up. Some plaster falls. Windows broken
6	Felt by all. Some panic. Trees sway. Some damage from falling objects
7	General alarm. Difficult to stand. Furniture broken
8	Panic. Branches broken off trees. Weak buildings fall down
9	Panic. Most buildings damaged. Underground pipes broken. Cracks appear in the ground
10	Panic. Many buildings fall down. Large landslides. Ground badly cracked. Rails bent
11	Panic. Most buildings fall down. Very wide cracks in the ground
12	Superpanic. Damage nearly total. Waves seen on the ground

Fig. 46
A scale of earthquake intensities

Fig. 47
Wreckage of a school viewed from the playground—one of the effects of the Great Alaskan Earthquake, 1964

We do occasionally get earthquakes in Britain. Some are natural but others may be caused by mining. None of these has ever been really serious. But in some parts of the world, violent earthquakes are only too well-known (Fig. 47) and they can kill thousands of people in a very short time.

Look at the extracts (Figs. 48 and 49) which describe two earthquakes, one in England and one in the U.S.A.

- Give each of these earthquakes a number on the scale in Fig. 46

A study of the distribution of recent earthquakes and volcanoes provides vital clues for unravelling the history of the earth.

(from the Guardian 4/11/76)

Earthquake shakes new tower blocks

Merseyside could experience a number of after-shocks after the earth tremor which yesterday rocked a 180-square-kilometre area in the North-west.

Widnes appeared to be the epicentre and the 400 staff at Halton District Council's eight-storey steel and glass HQ were evacuated. The building moved sideways by fifteen centimetres in 25 seconds.

A man on the third floor was thrown several feet across the floor 'as though someone had kicked my chair from behind'.

Walls quivered, shelves shifted, even ceilings moved. On the top floor, draughtsmen said their drawing boards moved when the building 'wriggled'.

In another Widnes office, a solicitor's receptionist stared up at her cup of tea sliding to the edge of her desk.

There was still doubt last night whether a three seconds tremor at the British Rail HQ in Crewe yesterday morning was linked with the Merseyside tremor. Crewe is nearer Stoke-on-Trent, which has had more than 100 earth tremors in the past 18 months, some of which have been blamed on mining operations.

Fig. 48

(from Stories of Famous Natural Disasters, Richard Garnett – published by Arthur Barker, 1976)

Doomsday at San Fransisco 1906

The first tremors knocked over lamps, shortcircuited electric cables, fractured gas mains, and before very long they had created thirty major fires. This was bad enough but the situation was dramatically worsened at 9.30 a.m. when the city's gas works blew up.

At first, it was the poorer quarter of San Francisco which suffered most. Blocks of tenements crumbled down on to the streets. The shock itself only lasted for three minutes. But it was sufficient to dispose of the City Hall which had been built at a cost of £1 400 000. It knocked over the Valencia Hotel, leaving seventy-five people buried beneath the debris of its five storeys. At another lodging house, between seventy-five and eighty people were entombed beneath blazing ruins.

Fig. 49

Fig. 50 shows the positions of some of the main earthquakes recorded in the last few years. You will see that these are often grouped in quite narrow zones, and earthquakes are rare or unknown in other places.

Put a piece of tracing paper over Fig. 50 and mark each corner.

- On your trace, draw an earthquake zone (i) in the middle of the Atlantic Ocean, (ii) that almost rings the whole of the Pacific Ocean, and (iii) that runs from Europe (Italy) to the Himalayas.

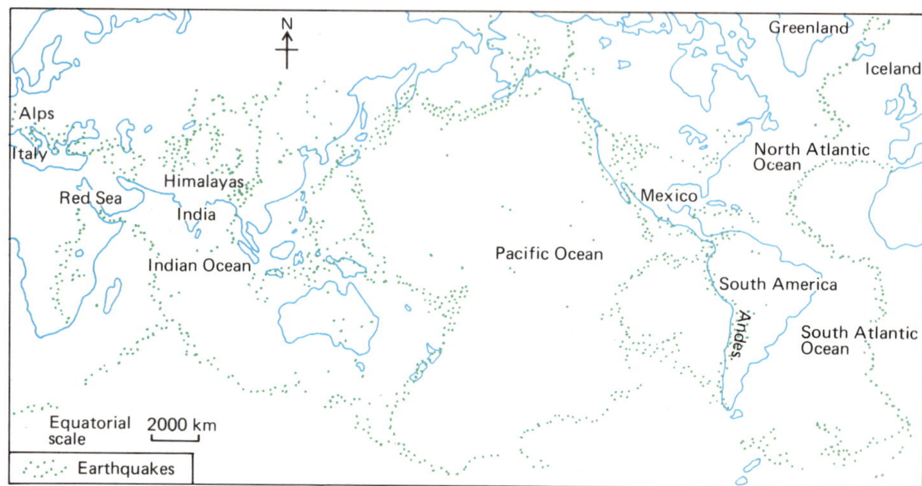

Fig. 50
The world's main earthquake zones

Now look at the places where there are active volcanoes on the surface (Fig. 51).

Put your trace from Fig. 50 over Fig. 51.

- How well do the positions of the active volcanoes match the earthquake zones?

The match is even better than it looks because Fig. 51 doesn't show the submarine volcanoes which are erupting all along the earthquake zones in the middle of the oceans. Only a few of these, like the Icelandic volcanoes, actually reach the surface of the sea and form islands.

This link between volcanoes and earthquakes is an important one. Both volcanoes and earthquakes build mountains, though unlike volcanoes, one earthquake doesn't make one mountain.

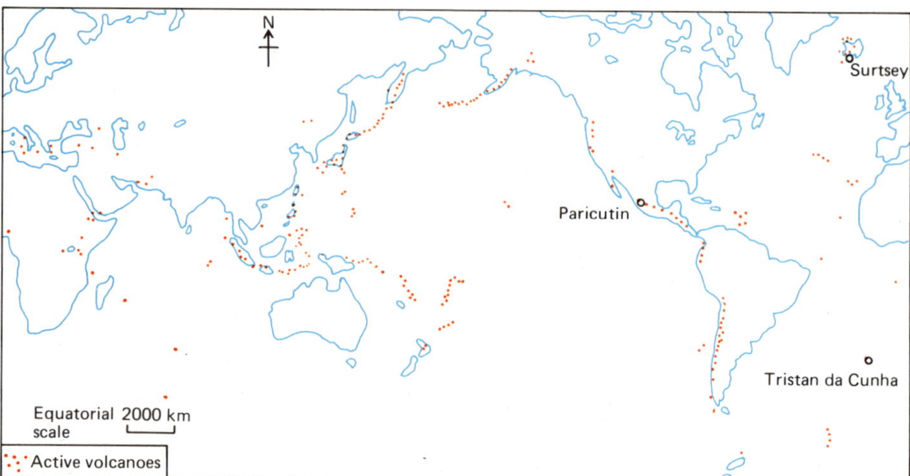

Fig. 51
The world's main active volcanoes

Most scientists now explain how earthquakes occur and how mountains are formed using a very unlikely sounding idea. They believe that the crust and the top of the mantle are broken into several large 'plates' that can move very slowly over the rest of the mantle. Some of these plates have only oceanic crust but most have both oceanic crust and continental crust. So as the plates move, the continents are rafted across the surface of the earth.

If this idea is right, it must mean that the pattern of land and sea has changed a lot through the earth's history. For example, it seems likely that there was no Atlantic Ocean or Indian Ocean 200 million years ago, and that all the continents were joined together.

One clue that helps us to build up the story of the moving plates is found in the basalt lavas erupted by submarine volcanoes. The lavas come from ridges standing well above the general level of the sea floor. And it seems that they don't build up layers one on top of the other. Instead, the old crust moves sideways and opens up a gap which is filled by the new crust (Fig. 52).

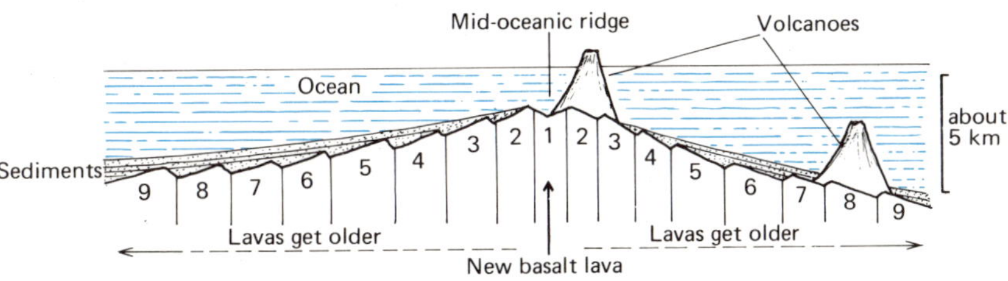

Fig. 52
The pattern of lavas and volcanoes under the sea

So some of the oceans are getting wider as more and more new crust is formed on either side of the ridges. The rate at which this has happened for the Atlantic and Pacific oceans over the last few million years is given in the table (Fig. 53).

Ocean	Rate of widening in cm per year
North Atlantic	2
South Atlantic	4
Pacific	9

Fig. 53
Widening oceans

You have already marked on your trace the earthquake zone along the mid-Atlantic ridge where new crust is being formed.

- Put your trace back over Fig. 50 and now mark the ridge that runs
  - (i) from the Red Sea through the Indian Ocean,
  - (ii) south from the coast of Mexico into the Pacific Ocean.
- Assuming that the oceanic crust spreads outwards on both sides at right angles to the ridges, mark with arrows the directions of widening for each of these three ridges.

The North Atlantic is still getting wider today. But there was a time when North America, Greenland and Europe were joined into one continent. How long ago was this?

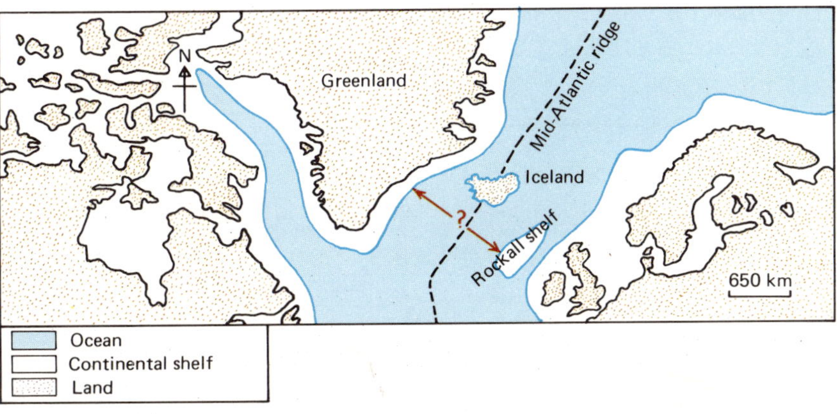

Fig. 54
The widening of the North Atlantic Ocean

Fig. 54 shows the distance from the edge of the Greenland continental shelf across the mid-Atlantic ridge to the edge of the continental shelf around Rockall. This is the width of the true ocean because the continental shelves are made of continental crust, not oceanic crust. The oceanic crust only begins beyond the continental shelves.

Suppose that the rate of widening of the North Atlantic has stayed the same over all the time the ocean was forming.

- Work out roughly how long ago the opening of the ocean must have begun.

This is about the time of the great burst of igneous activity in western Scotland and Northern Ireland (see Further study 5: *Rock cycles in the past*). The area must then have been very much closer to the middle of the new ocean than it is now.

If new crust is forming on the ocean floor, then old crust must be 'disappearing' somewhere else. Look at the arrows you have marked on your trace. Two of them should be pointing towards one another across South America, one from the ridge in the Pacific Ocean and one from the ridge in the Atlantic Ocean.

The South American plate, which carries South America on its back, meets the plate in the east of the Pacific Ocean at the earthquake zone on the west coast of South America. Here (Fig. 55) the denser oceanic crust goes under the lighter continental crust. The friction caused by one plate moving under another results in earthquakes. And in some places, molten material reaches the surface in volcanoes.

The earthquake zones in the middle of continents have to be explained in another way. Think of the opposite of what is happening in the Atlantic—that an ocean is shrinking (Fig. 56), not getting wider. The two continents carried on the backs of the plates must finally collide.

When they do, the continental crust has too low a density for it to sink like oceanic crust. Instead, it is squeezed by huge forces so that the rocks are crumpled (folded) and sometimes broken (faulted) to cause earthquakes.

The rocks that are deeply buried during these events are at a high temperature and under a very high pressure. This may begin to change them into completely different rocks called metamorphic rocks. The process is called metamorphism.

The land is then uplifted to form high mountains.

If you imagine one land mass in Fig. 56 to be Europe (minus Italy) and the other to be Italy, then this is the collision that pushed up the Alps in Tertiary times (page 79). To make the Himalayas, the plate carrying India must have collided with the plate carrying the rest of Asia.

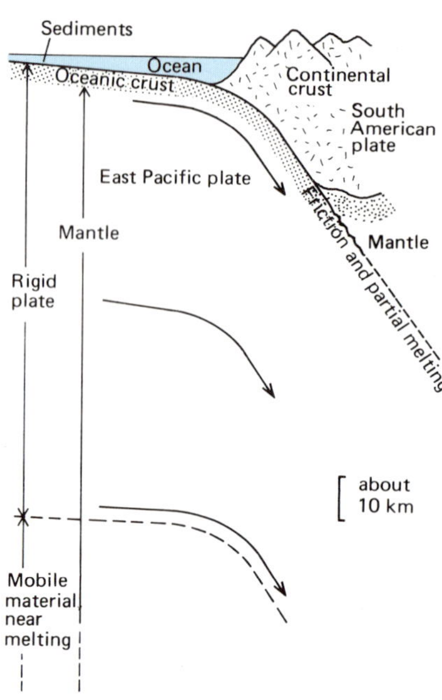

Fig. 55
One kind of plate margin where oceanic crust runs under continental crust

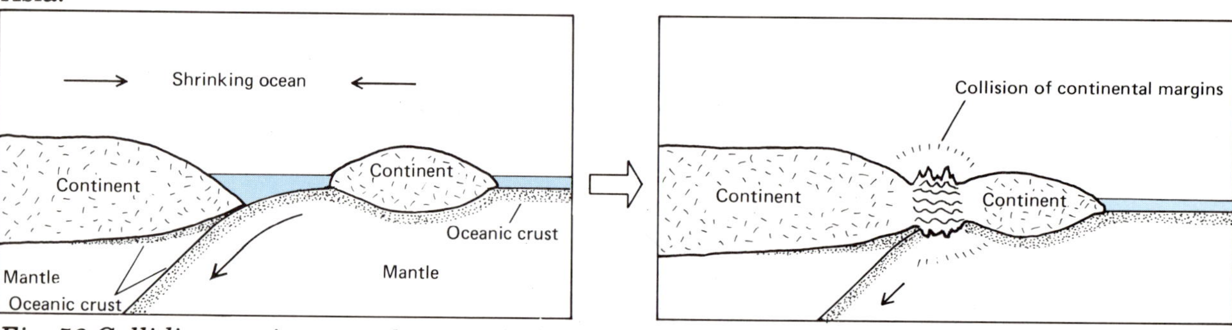

Fig. 56 Colliding continents and mountain-building

In this way, today's jigsaw of continents and oceans can be fitted together. This has been done in Fig. 57 for the British Isles over the last 500 million years. Both the collisions shown in the diagrams pushed up high mountains which have long since been eroded to their 'stumps'.

Activity 8 shows one of the things that happens to layers of rock during a continental collision.

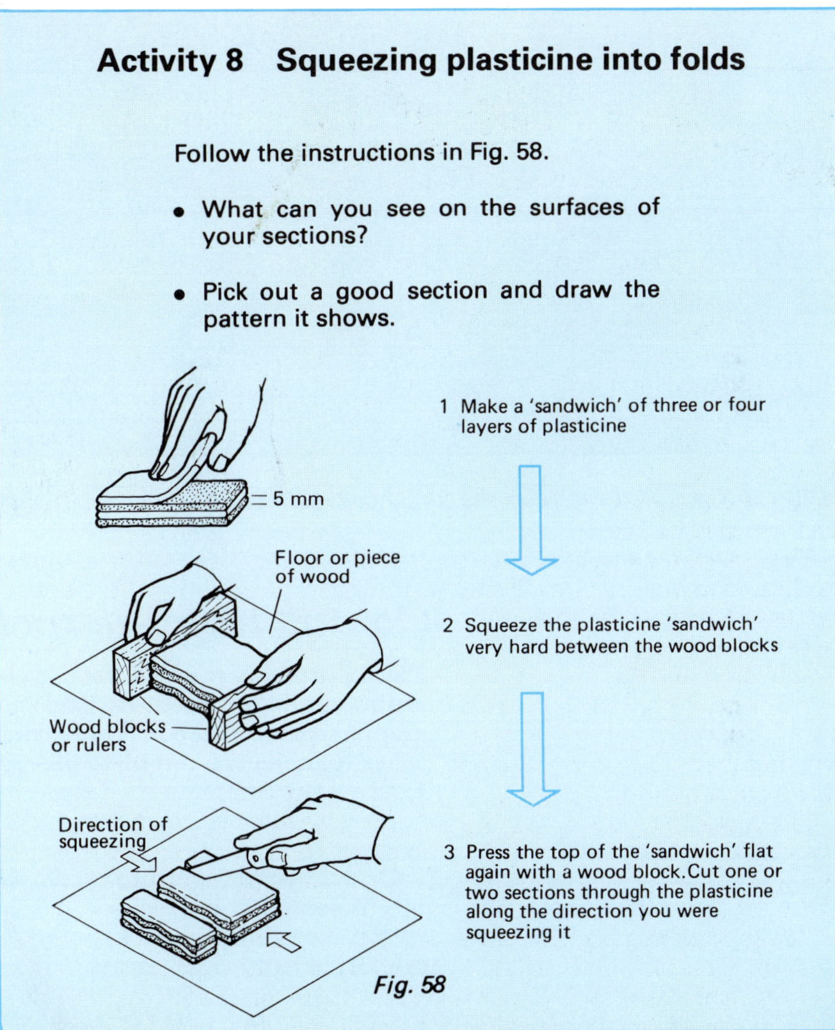

Fig. 58

You now know the story of how sediments formed under water can end up as rocks high above sea level. The rocks may be sedimentary or metamorphic and they may be very crumpled like those in Fig. 59. But they may also end up only slightly crumpled if they have been uplifted but not squeezed very much.

60 m.y. ago
North American continent
Widening ocean
European continent
Opening of North Atlantic so separating North America from Europe

300 m.y ago
Final closing of ocean. Continental collision results in mountain-building

350 m.y ago
North American – North European continent
Shrinking ocean
Continent with Spain and Africa
Most of Britain separated from south-west England by a shrinking ocean

400 m.y ago
North American – North European continent
Final closing of ocean. Continental collision results in mountain- building

500 m.y ago
North American continent
Shrinking ocean
North European continent
Northern Britain part of North American continent. Separated from southern Britain by a shrinking ocean. Some mountains in northern Britain already forming

Ocean
Land and continental shelf
High mountains

Fig. 57
Some important events in the geological history of Britain

Fig. 59 Crumpled rocks on the foreshore near Saunton, north Devon

Completing the rock cycle

Fig. 44 is one part of the rock cycle. The change from thick layers of sediments to sedimentary and metamorphic rocks in mountains is another part (Fig. 60). These two parts can be plugged into each other to make the complete picture.

- Draw the complete rock cycle.

As you can see, this large cycle is made from several smaller cycles.

- Draw the two separate cycles which show how mountains of
  (i) sedimentary rocks
  (ii) metamorphic rocks

are formed and worn away.

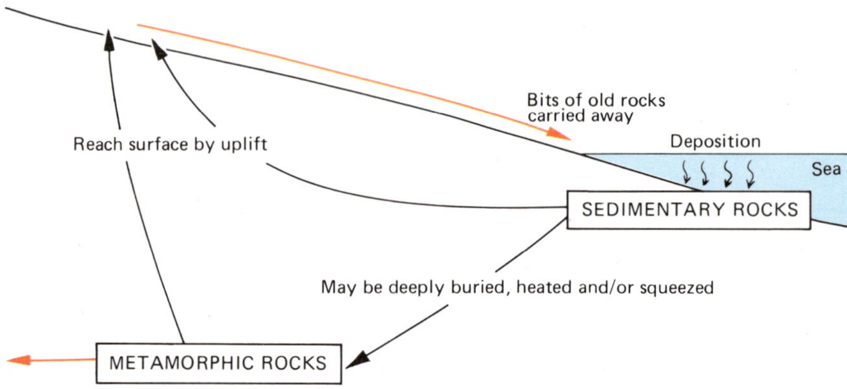

Fig. 60
The part of the rock cycle that concerns metamorphic rocks and mountain-building

Recognizing minerals and rocks

The main theme of this book is that the history of the earth can be pieced together by matching some processes that are taking place now with evidence from rocks. But to get this evidence, you have first to know how to study rocks and how to recognize them. This section will help you to do these things.

Minerals

A mineral is a single substance that can be found in nature. Different samples of the same mineral look very alike (except perhaps in colour) and they also behave in the same way when they are tested.

Just as for the solids on the shelves of a chemistry laboratory, each mineral can usually be given a chemical name and formula.

Not all minerals are completely pure when dug up. The impurities in them may affect their colour. So halite (rock salt) is often brown because it contains some clay, but pure halite is colourless or white.

Everyone would like to collect samples of minerals like those in Fig. 61. You could use their shapes to identify them.

(a) Calcite crystals;

(b) Galena crystals;

(c) Haematite (kidney iron ore)

Fig. 61

But you won't always have such perfect examples to look at. Good crystals are only formed under special conditions. Instead, you will have to rely on other observations to make an identification.

Hardness is a very useful test for a mineral. An order of hardness, called Mohs' scale, is given in Fig. 62. For each hardness number, one mineral is used as the standard.

The minerals get harder to scratch as you go up the scale. A mineral will scratch all those minerals that are below it in the scale and will not scratch those above it.

The hardness of three everyday materials are also shown in Fig. 62. But you must bear in mind that both finger nails and the steel in knife blades can vary in hardness a little.

- Which mineral can be scratched by a 10p coin but not by a finger nail?
- Which minerals can be scratched by a penknife blade but not by a 10p coin?

The way the mineral splits (cleaves) is also a useful test. In Fig. 63 a lump of calcite is being hit in the direction of the cleavage with a sharp implement. You end up with smaller bits of calcite of a definite shape, and this shape is made by three sets of cleavage faces.

The angles that the cleavage faces make with one another are closely linked to the shape of the crystals. In the case of halite, the cleavage fragments look exactly the same as the crystals. But for calcite, they look very different.

Hardness	Mineral	
10	Diamond	
9	Corundum	
8	Topaz	
7	Quartz	
6	Feldspar (Orthoclase)	
		← a penknife blade (Hardness 5½)
5	Apatite	
4	Fluorite	
		← a 10p. coin (Hardness 3½)
3	Calcite	
		← a finger nail (Hardness 2-2½)
2	Gypsum	
1	Talc	

Fig. 62
Mohs' Scale of Hardness

Activity 9

For this activity, you should choose two minerals that look alike, perhaps because they are both colourless or white. What you have to do is to find ways of telling them apart. Carry out each of these tests on both minerals and put your results in a table with the headings shown in Fig. 64.

Crystal shape and cleavage

Use a hand lens to look for flat faces that could be crystal faces or cleavage faces, or both.

- Draw any crystals with a definite shape that you can see.
- Does the mineral show any cleavage faces? If so, how many sets of faces are there and what angles are they to one another?
- Does the mineral show any irregular breaks?

Hardness

See if you can scratch the mineral with your finger nail, a 10p. coin or a penknife. Each time you try this, rub the 'scratch' with a wet finger and look at the mark under a hand lens. Decide carefully whether it is a proper scratch and not just a few loose grains of the mineral or a grey mark made by the coin or the knife.

- Use Fig. 62 to decide roughly where the two minerals come in Mohs' scale.
- Which mineral is likely to be able to scratch the other? Test your idea out to see if you are right.

Colour and lustre

- What colour is the mineral?
- Is it transparent (can be seen through), translucent (lets light through but can't be seen through), or opaque (can't let light through).
  Hold up the mineral to the light so that you can see its lustre (shine).
- Which one of these lustres does the mineral have?
  dull glassy
  metallic none of these

Streak

If the mineral can be scratched, look at its powder.

- What is the colour of this powder?

To confirm this colour, make a streak with the mineral by scratching a sharp edge across a tile of white unglazed porcelain. For some minerals, the streak is the same colour as the whole mineral; for others, it is a different colour.

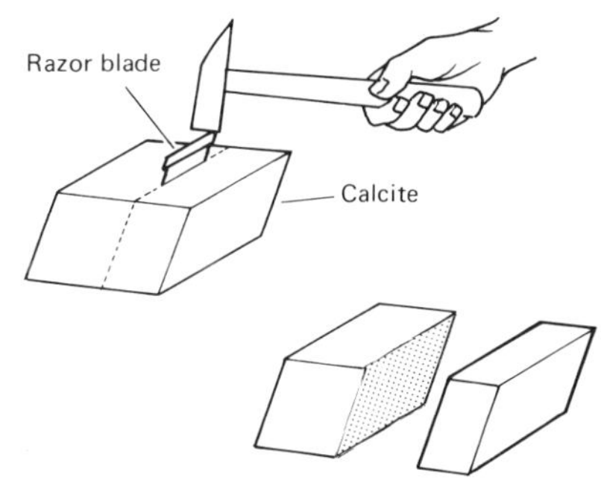

Fig. 63

Some minerals don't split in a regular way: they have no cleavage. Despite this, they may still break in a way that can be used to identify them. Quartz is a good example: it breaks to form a curved and ribbed surface which looks like some kinds of sea-shell.

Activity 9 shows how you can describe minerals and how you can tell two minerals from one another. But you might also want to know how to identify unknown minerals.

One way of doing this is by a chemical analysis. Adding dilute acid to the mineral is part of such an analysis but there are many other chemical tests besides this.

Activity 10 gives you a quicker way of identifying some common minerals using some of the tests in Activity 9.

Comparing two minerals

Mineral	Crystal shape and cleavage	Hardness	Colour and lustre	Streak	Reaction with acid	Relative density
A						
B						

Fig. 64

Reaction with acid

Follow the instructions in Fig. 65 using some drops of dilute hydrochloric acid.

- Does the mineral 'fizz' (effervesce)?

Relative density

Find the relative density of the mineral using the method given in Activity 7 (page 16).

- Which of the two is the denser mineral?

Fig. 65

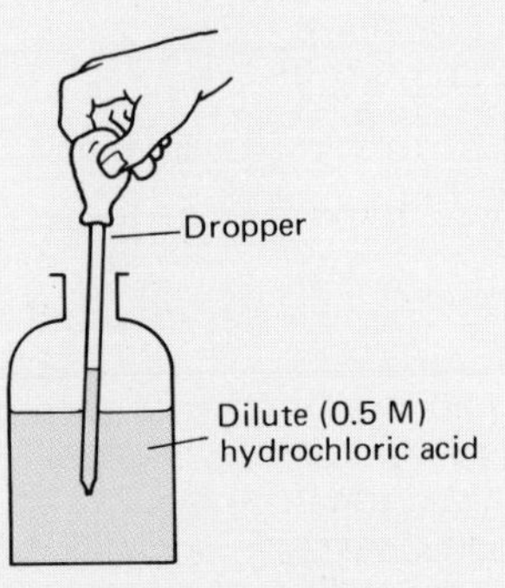

1 Take up a small amount of the acid into a dropper

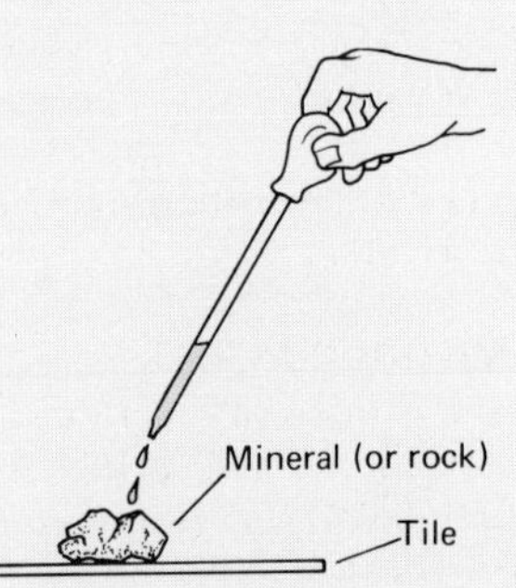

2 Add 1 or 2 drops of the acid to the mineral (or rock)

Activity 10 Using a flowchart to identify some minerals

Follow the flowchart (Fig. 66) with your mineral until you reach the end of one of the 'trails'.

Because the flowchart only contains a few minerals, your own mineral may turn out not to be the one named there. So you should check this by looking for more details in Fig. 67. If your mineral doesn't match any of the descriptions very well, you may need the help of a textbook on minerals.

- Can you identify your mineral for certain?

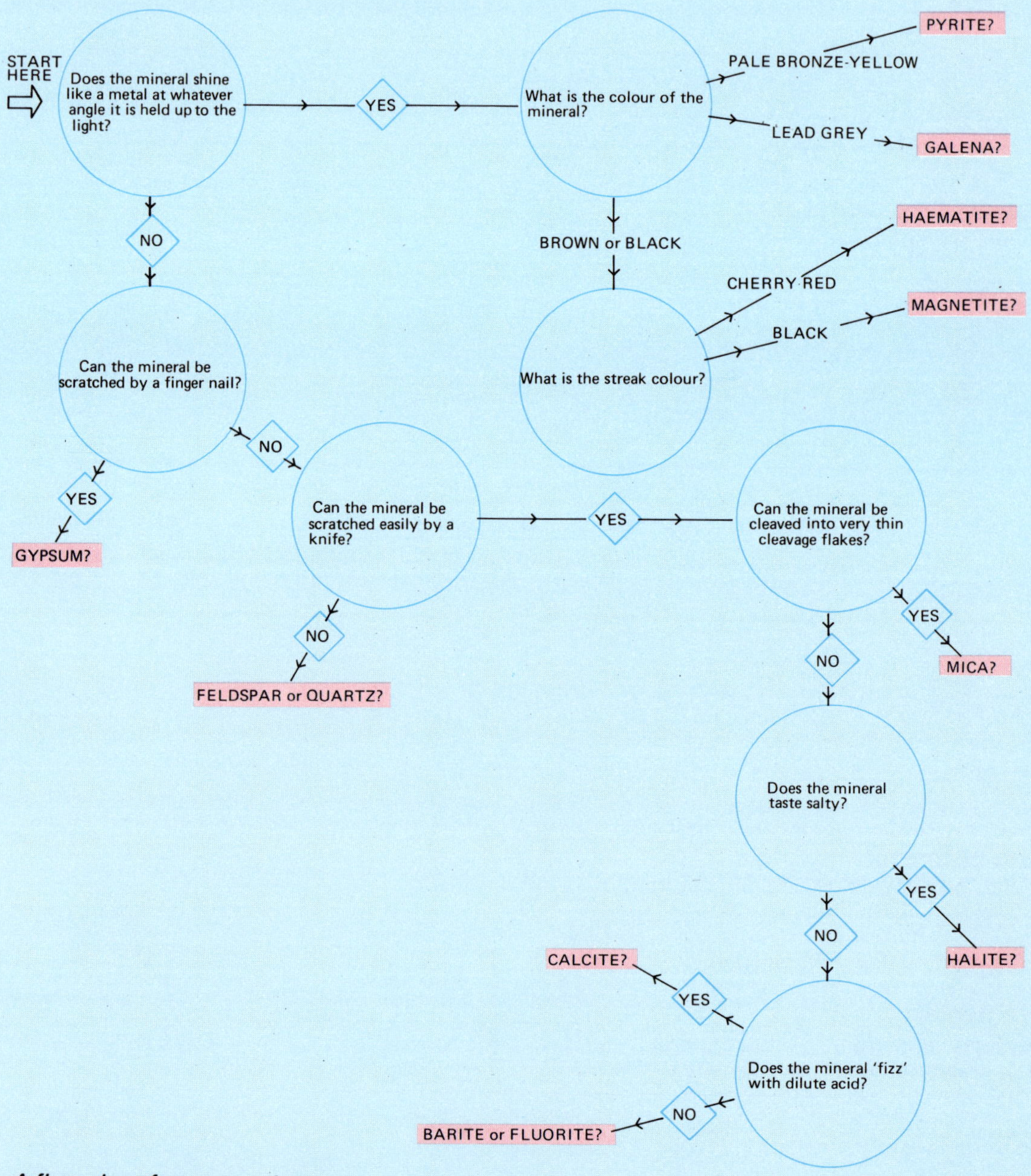

Fig. 66 A flow chart for some minerals

Mineral	Chemical formula	Colour	Hardness	Relative density	Other properties
Anhydrite	$CaSO_4$	White	3½	2.9	Good cleavage, often found in fibres or grains.
Barite	$BaSO_4$	Colourless, white, pink, blue	3	4.5	Crystals have broad flat surfaces, glassy lustre
Calcite	$CaCO_3$	Colourless, white	3	2.7	Good cleavage (forms rhombs—Fig.63), 'fizzes' with dilute acid
Feldspar	e.g. $KAlSi_3O_8$	White, pink	6	2.7	Good cleavage
Fluorite	CaF_2	Colourless, purple, yellow	4	3.2	Cubic crystals (like halite), glassy lustre
Galena	PbS	Lead grey	2½	7.5	Cubic crystals
Gypsum	$CaSO_4.2H_2O$	Colourless, white	2	2.3	Good cleavage, sometimes found in fibres or grains.
Haematite	Fe_2O_3	Red/black, grey	6	5.2	Metallic lustre, cherry red streak, becomes magnetic on heating
Halite (rock salt)	$NaCl$	Colourless, white, sometimes brown	2½	2.2	Cubic crystals (Fig.23), good cleavage (forms cubes), glassy lustre, tastes salty
Magnetite	Fe_3O_4	Iron-black	6	5.2	Metallic lustre, black streak, strongly magnetic
Mica	e.g. $KAl_3Si_3O_{10}F_3$	White, brown, black	2½	2.9	Good cleavage (forms very thin flakes), glints when held up to the light at a certain angle
Pyrite	FeS_2	Pale bronze -yellow	6½	5.0	Crystal faces often covered in parallel lines (striations), metallic lustre, greenish black streak
Quartz	SiO_2	Colourless, white	7	2.7	Glassy lustre, no cleavage — breaks to form curved, ribbed surfaces
Sphalerite	ZnS	Black/ brown	4	4.1	Good cleavage, white to reddish brown streak, no metallic lustre

There are several different kinds of feldspar and mica with slightly different chemical formulae and properties

Note The hardness and relative density of a mineral may vary a little from sample to sample

Fig. 67 The properties of some common minerals

Rocks

Rocks are made from grains or crystals of minerals. There are usually two or more different minerals in each rock, though a few rocks, such as chalk, may have grains of one mineral only.

Fig. 70 shows the three main kinds of rock you can find. It isn't always easy to tell whether a rock is made from grains or crystals. But this is the first thing you should try to find out when you are describing or identifying a rock sample.

Rocks made from crystals are often tougher than rocks made from grains—they are harder to break up into smaller bits. This is because the crystals are strongly bound together and any breaks have to form within the crystals themselves.

The grains in many sedimentary rocks are only bound together loosely and the cement between the grains is easily broken up. It is

Activity 11

For this activity, you should choose two rocks. One could be sedimentary and the other igneous or metamorphic. Carry out each of these tests on both rocks and put your results in a table with the headings shown in Fig. 68.

Rock	Texture	Toughness	Colour
X			
Y			

Fig. 68

Texture

Look at the rock without a hand lens and then with a hand lens.

- Say whether the rock is coarse-grained (grains or crystals can be seen without a hand lens), medium-grained (grains or crystals can be seen only with a hand lens) or fine-grained (grains or crystals can't be seen even with a hand lens).
- Do the particles look like sharp-edged crystals or more rounded grains?
- Can you see any tiny surfaces that glint in the light? (The glinting is likely to be due to crystals and not grains.)

Toughness

Choose a part of the rock that is 'fresh' and not a part which has been altered (weathered) by the action of the air. Follow the instructions in Fig. 69. Look at the broken-up bits with a hand lens.

- Is the rock made from grains or crystals?
- How tough is the rock?

Colour

- Is the rock one colour or are there crystals or grains of different colours?
- If there are different colours, roughly what fraction of the rock is made from each of the colours?
- Can you identify any of the minerals in the rock?

Layering

- Can you see any bedding planes in the rock?
- Is there any slaty cleavage?

only when the cement is very strong and the grains are squeezed very closely together that a rock made from grains might be as tough as a rock made from crystals.

To a geologist, a rock can vary from a tough granite or basalt used for roadstone to a loose sand or clay. The only loose material in the earth that doesn't count as rock is the soil which forms on the surface.

- What is the main difference between a soil formed on a loose sand or clay and the sand or clay itself?

Another way to tell a sedimentary rock from an igneous or metamorphic rock is to look for layering. But there are problems here. Bedding planes are often a long way apart and a small sample of a sedimentary rock may have been taken from a place where there are none. The absence of bedding planes doesn't prove that the rock must be igneous or metamorphic. In any case, volcanic igneous rocks are often layered.

Comparing two rocks

ering	Reaction with acid	Fossils	Relative density

- If the rock is made from crystals, are the crystals arranged in light- and dark-coloured bands? Or are there mica crystals arranged in a particular direction?

Reaction with acid
Follow the instructions in Fig. 65.

- Does the rock 'fizz' (effervesce)?

Fossils
- Are there any fossils in the rock?

Relative density
If the rock sample feels heavy for its size, find its relative density (Fig. 37).

- Is the relative density much higher than the average for rocks in the earth's crust?

1 Rub the rock between your fingers

Go on to stage 2 if the rock doesn't crumble in stage 1

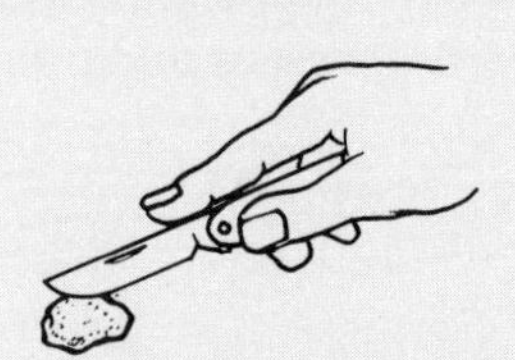

2 Scrape the rock firmly with a penknife

Go on to stage 3 if the rock doesn't crumble in stage 2

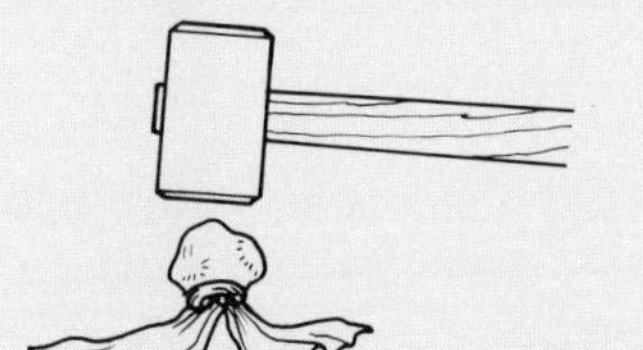

3 Wrap the rock in a cloth and hit it with a hammer.

Fig. 69

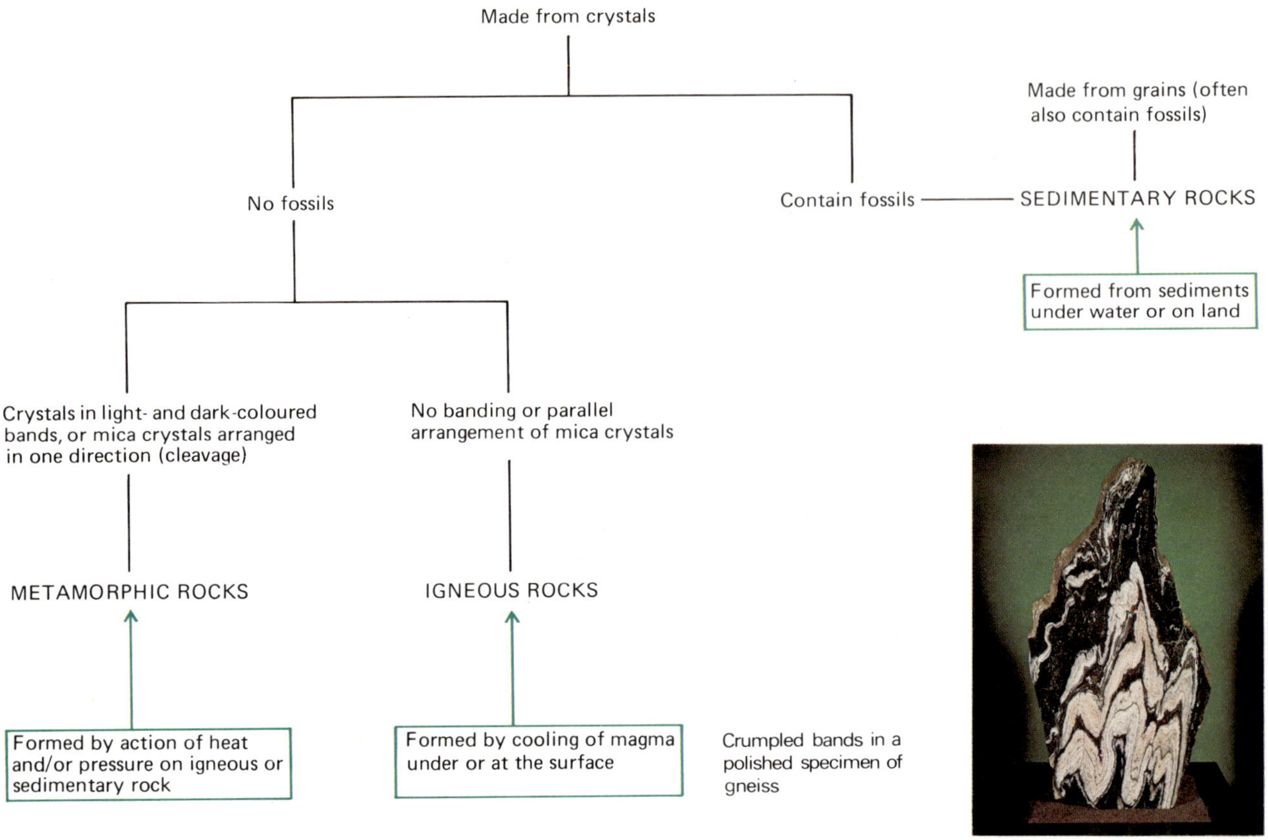

Made from crystals
Made from grains (often also contain fossils)
No fossils
Contain fossils
SEDIMENTARY ROCKS
Formed from sediments under water or on land
Crystals in light- and dark-coloured bands, or mica crystals arranged in one direction (cleavage)
No banding or parallel arrangement of mica crystals
METAMORPHIC ROCKS
IGNEOUS ROCKS
Formed by action of heat and/or pressure on igneous or sedimentary rock
Formed by cooling of magma under or at the surface

Crumpled bands in a polished specimen of gneiss

Fig. 70
A simple way of classifying rocks

Also, you can see a kind of 'layering' in some metamorphic rocks (see the photo in Fig. 70). The 'layering' in slates (Fig. 71) is cleavage and not bedding. Just like a lot of minerals, slate can be split very easily along the lines of cleavage. Slaty cleavage is formed when fine-grained rocks such as shales are squeezed sideways by strong forces during mountain-building movements.

Fig. 71
Slaty cleavage. The quarry is at Llanberis in Gwynedd

Cleavage in metamorphic rock
Direction of pressure
Folded layers of former sedimentary rock

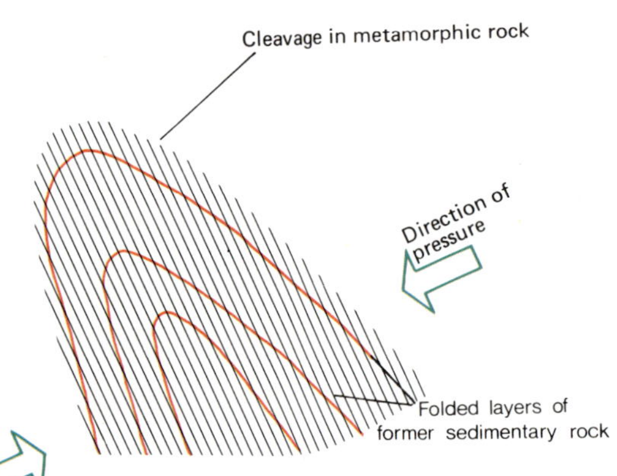

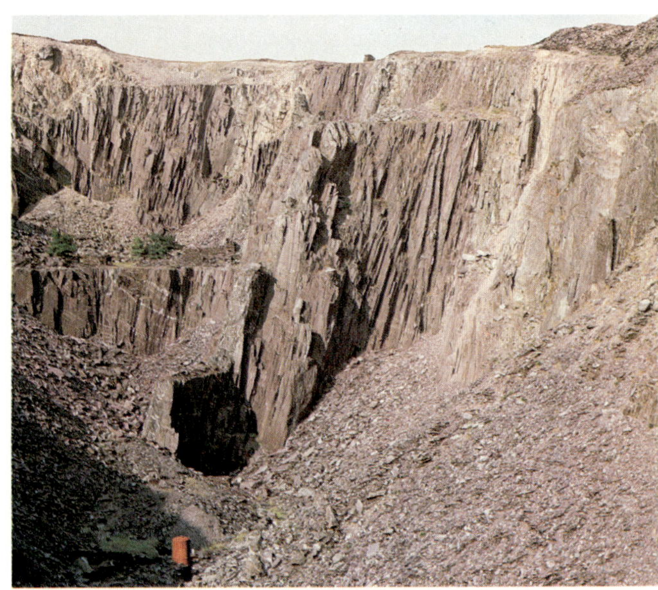

Activity 12 will help you to identify some common rocks using the tests in Activity 11.

Activity 12 Using a flowchart and some keys to identify some rocks

One of the questions in Fig. 72 may be rather difficult to answer—can you see crystals, grains or pebbles with or without a hand lens? If you do find it difficult, you may have to look at some typical samples of each kind and so compare these with your unknown rock.

Notice that we often use the words 'fine-grained', 'medium-grained' and so on for both rocks made from grains and rocks made from crystals.

Follow the flowchart (Fig. 72) with your rock until you come to the end of one of the 'trails'. You may then have to go on to one of the keys (Figs. 73–76).

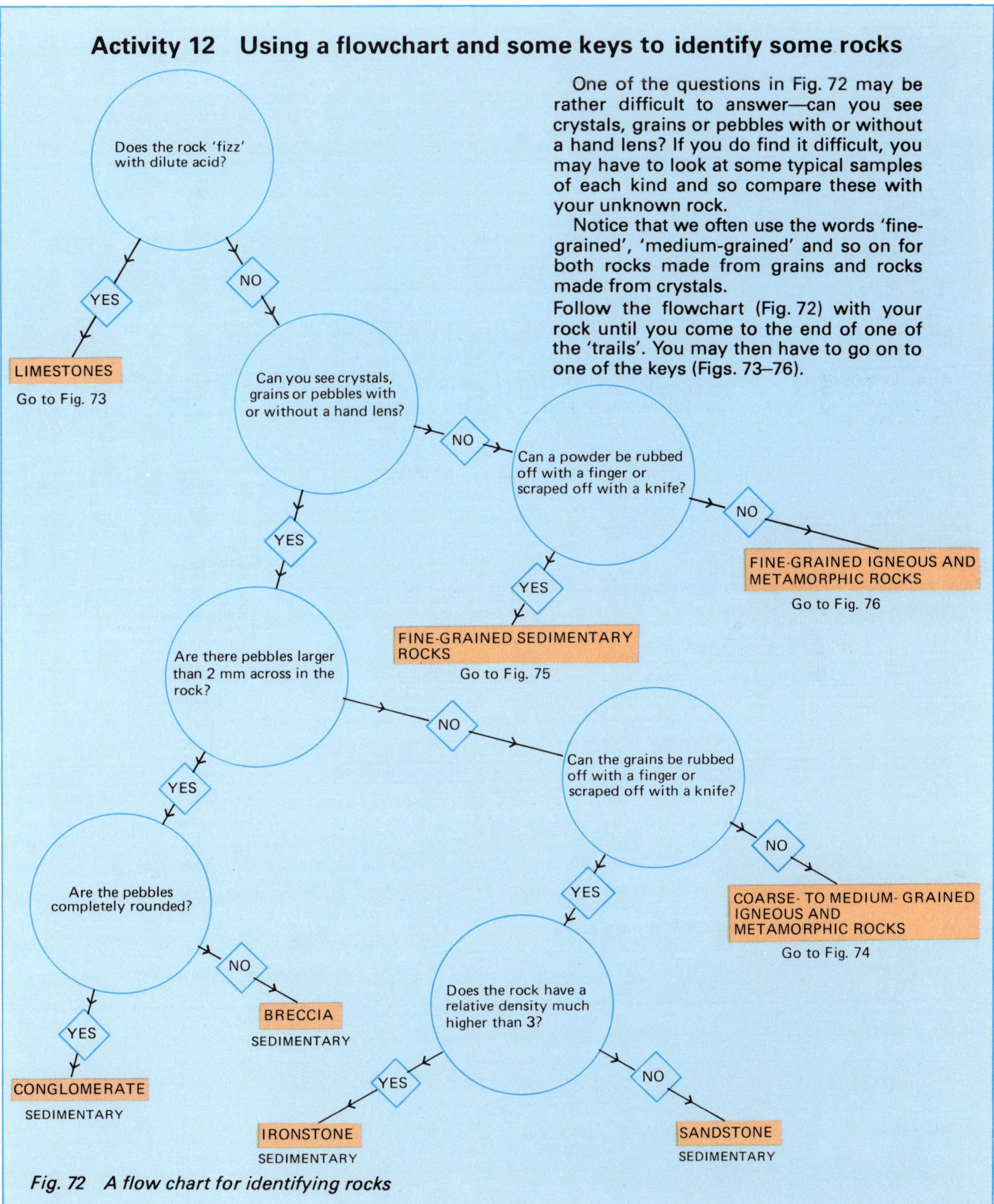

Fig. 72 A flow chart for identifying rocks

LIMESTONES
Crumbles easily to a white powder
Has small egg-like grains
Has bits of calcite fossils
CHALK
SEDIMENTARY
OOLITIC LIMESTONE
SEDIMENTARY
SHELLY LIMESTONE
SEDIMENTARY

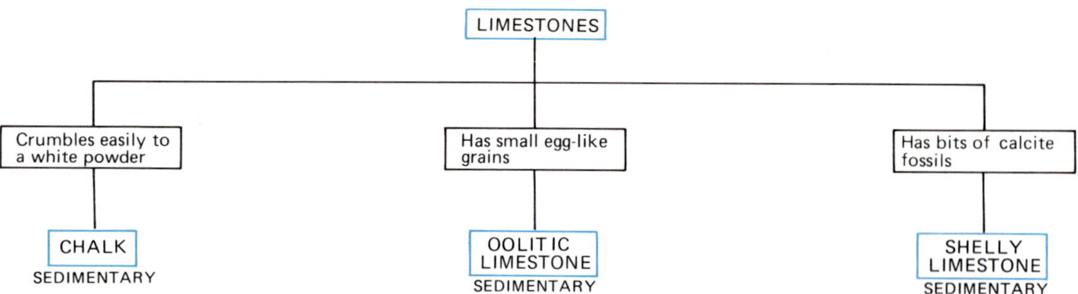

Fig. 73 A key for limestones

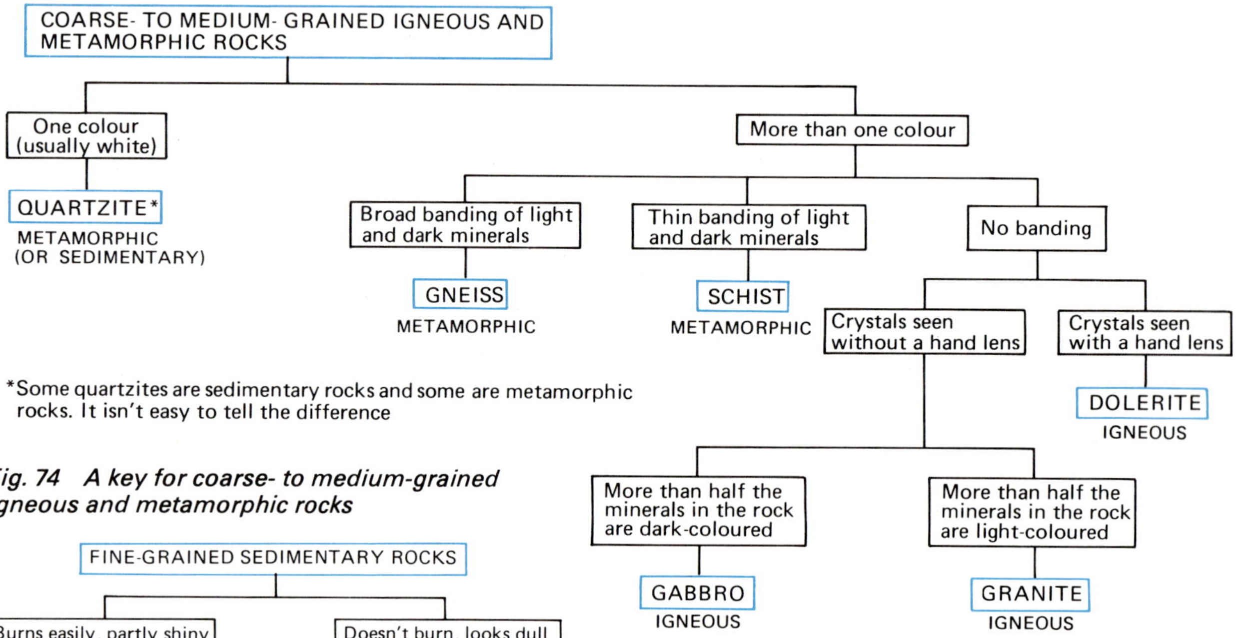

COARSE- TO MEDIUM- GRAINED IGNEOUS AND METAMORPHIC ROCKS
One colour (usually white)
More than one colour
QUARTZITE*
METAMORPHIC (OR SEDIMENTARY)
Broad banding of light and dark minerals
Thin banding of light and dark minerals
No banding
GNEISS
METAMORPHIC
SCHIST
METAMORPHIC
Crystals seen without a hand lens
Crystals seen with a hand lens
DOLERITE
IGNEOUS
More than half the minerals in the rock are dark-coloured
More than half the minerals in the rock are light-coloured
GABBRO
IGNEOUS
GRANITE
IGNEOUS

*Some quartzites are sedimentary rocks and some are metamorphic rocks. It isn't easy to tell the difference

Fig. 74 A key for coarse- to medium-grained igneous and metamorphic rocks

FINE-GRAINED SEDIMENTARY ROCKS
Burns easily, partly shiny
Doesn't burn, looks dull
COAL
SEDIMENTARY
Thin layering
No layering
SHALE
SEDIMENTARY
Can be moulded when wet
Can't be moulded when wet
CLAY
SEDIMENTARY
MUDSTONE
SEDIMENTARY

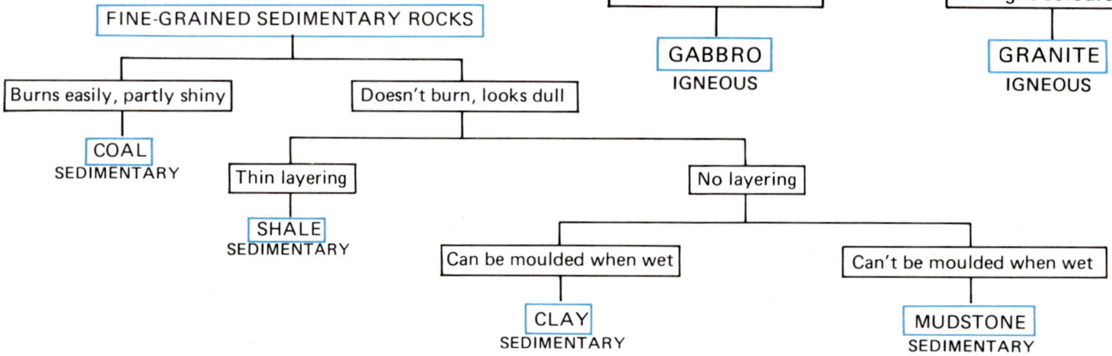

Fig. 75 A key for fine-grained sedimentary rocks

FINE-GRAINED IGNEOUS AND METAMORPHIC ROCKS
Cleavage, doesn't glint in the light
No cleavage, does glint in the light
SLATE
METAMORPHIC
Dark-coloured
Light-coloured
BASALT
IGNEOUS
RHYOLITE
IGNEOUS

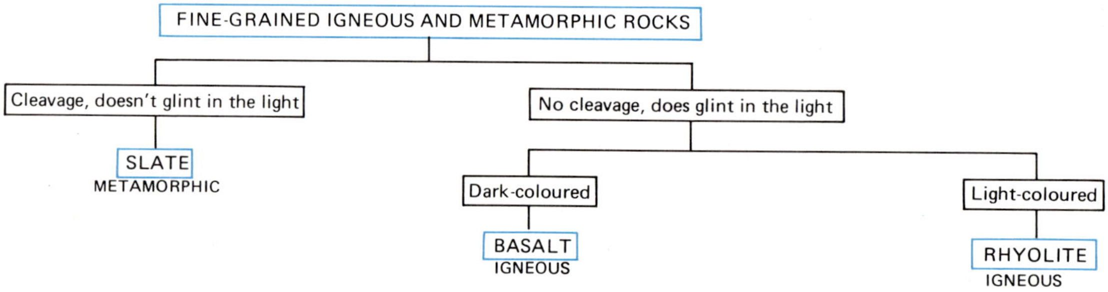

Fig. 76 A key for fine-grained igneous and metamorphic rocks

Life in the past

What are fossils?

Fossils are clues in the rocks that tell us something about life in the past. They are nearly always found in sedimentary rocks. Fig. 77 shows that these can vary from a whole insect preserved exactly as it was in life to just the tracks made by a worm.

Most fossils are bits of an animal or plant rather than the whole thing. Quite often, it is only some of the hard parts of an animal, such as the teeth, bones or shells, that are preserved. This is because the soft parts easily rot away. The most resistant parts of plants are the pollen grains, but the carbon prints of leaves and stems can also be found in some rocks.

- Why is it very rare to find a whole fossil insect?

Trace fossils are usually marks made by the activities of animals but they aren't the remains of the animals themselves. Some more trace fossils are shown in Figs. 78 and 79.

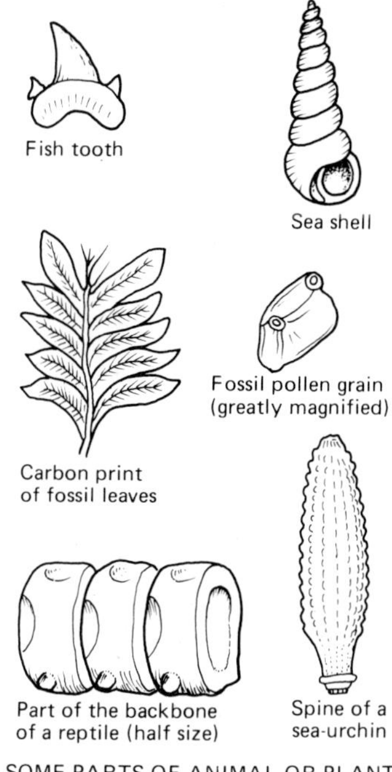

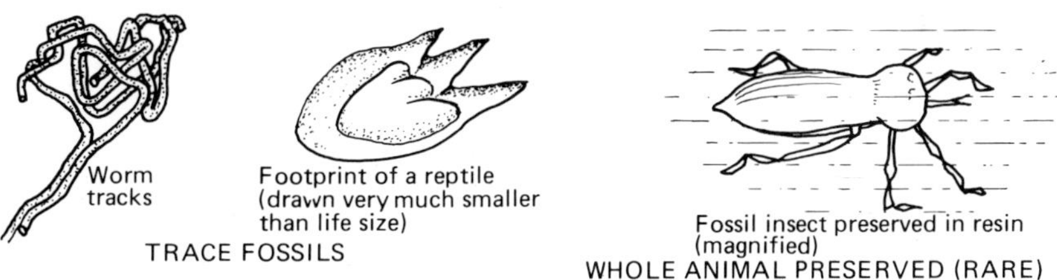

Fig. 77
Some examples of fossils

A piece of detective work

Just as when he studies rocks, a geologist has to be something of a detective when he looks at fossils. Sometimes the fossil is like an animal or plant still living today. But quite often, he has to deal with living forms which are now extinct. This makes it very hard for him to piece the bits of evidence together and so get a more complete picture of life in the past.

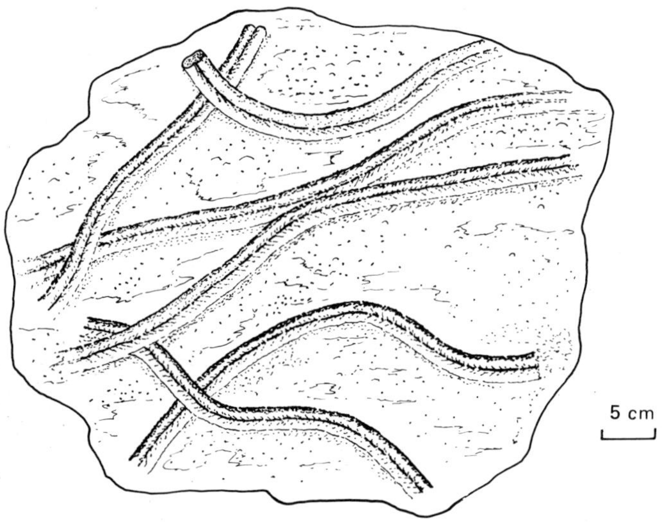

Fig. 78
Meandering furrows in a sedimentary rock from north Wales

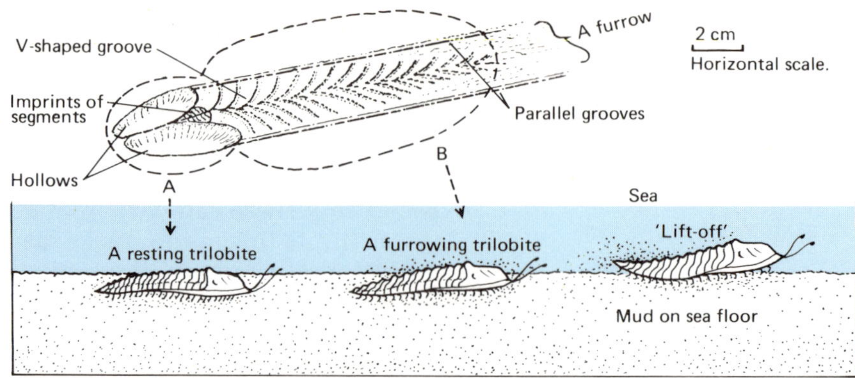

Fig. 79
Close-up of a furrow

What would you make of the marks shown in Fig. 78 which can be found on some very old rocks in north Wales? There is a clue to the answer in Fig. 79.

In the middle of the part of the marks labelled A there are imprints of segments of an animal like the one drawn in Fig. 80. This is a kind of trilobite and, like all trilobites, is now extinct. Trilobites can be found in some rocks along with fossil corals. Because modern corals live on the sea floor, it seems likely that trilobites also lived in the sea. Unlike corals, though, they had legs on either side of their bodies, and these might have been used for swimming or crawling. But at A, this particular trilobite may have been resting in the mud, partly buried.

Fig. 79 shows that part B of the marks may have been made when the trilobite started 'ploughing' through the mud. The legs had claws on the end which must have dug down under the body to make the V-shaped grooves. These grooves get smaller as the tribolite began to lift itself out of its furrow.

- How many claws might the trilobite have had on each leg?
- Use Figs. 79 and 80 to decide what might have made the two long parallel grooves at the edges of the track.

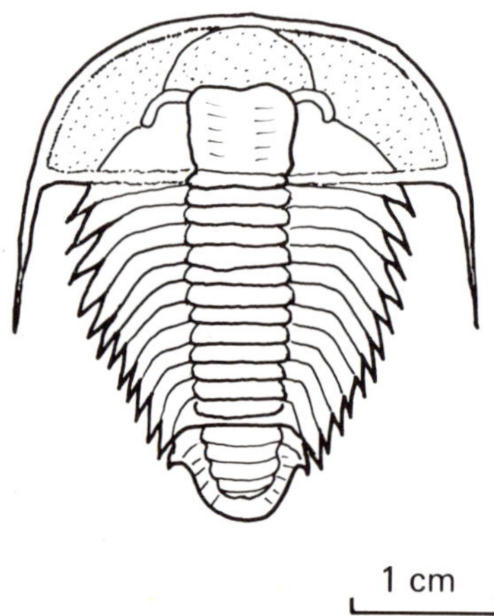

Fig. 80
A trilobite with spines

Part C of the track in Fig. 81 may have been made by the trilobite walking on the surface with its legs spread further apart rather than 'ploughing' through a furrow.

You now know something about some very short events in the life of an animal which lived over 500 million years ago! Look carefully at the difference in the track between part D and part C. Are you a good enough geological detective to finish off for yourself the story of this ancient trilobite's trip along the sea floor?

- Try to guess how the trilobite made the tracks in part D.

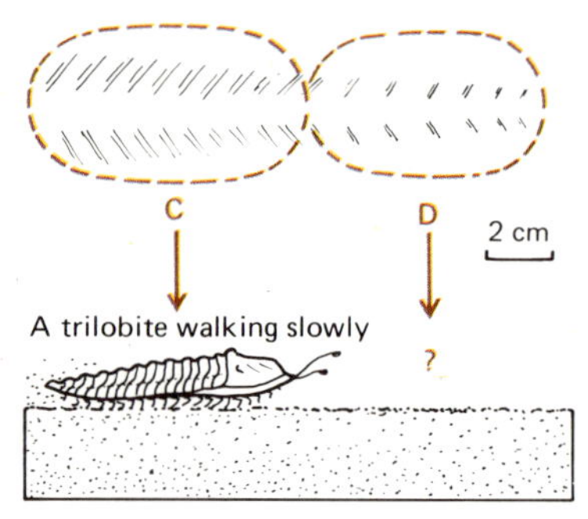

Fig. 81
More trilobite tracks

How are fossils formed?

Any living form is most likely to leave a fossil trace if it gets buried quickly beneath sediment before it can be destroyed.

- Is this quick burial more likely to take place on land or under water?
- Why are fossils of animals which lived under water much more common than fossils of land animals?
- Where today might you find the fossils of tomorrow?

Let us see how modern sea-shells can be fossilized (made into fossils). These shells 'fizz' with dilute acid because they are made from calcium carbonate. Some fossil shells do the same, but some don't.

The process of fossilization (forming fossils) can change the original shell a lot. You can find out more about the process by making your own artificial 'fossils'.

Activity 13 Making impressions and casts of shells

Follow the instructions in Fig. 82 using a modern shell. When you take the shell off the plasticine or clay, you will see an impression (a mould) of the shell. The plaster is then used to make a cast of the impression.

- How well does the cast show the markings on the shell?
- Is the detail as good as on the impression?

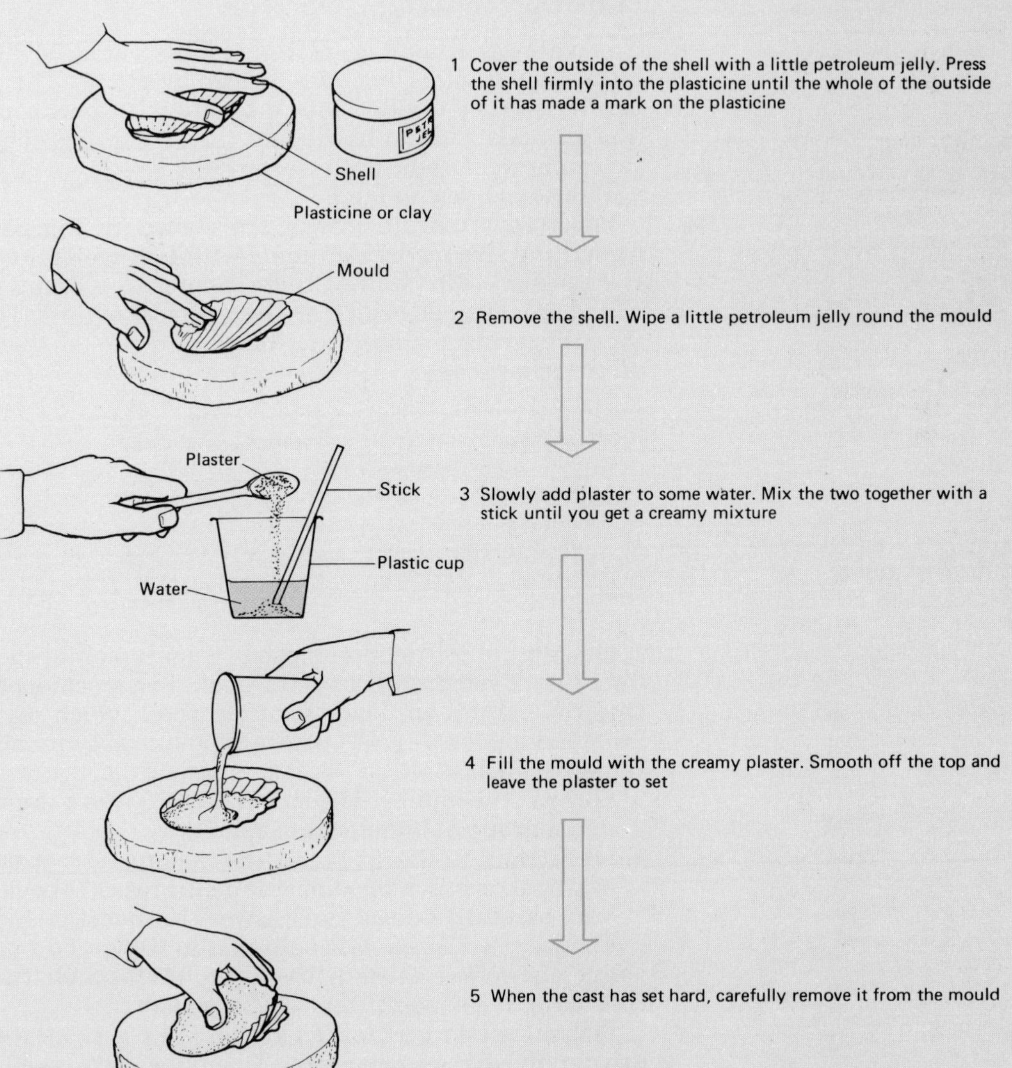

Fig. 82

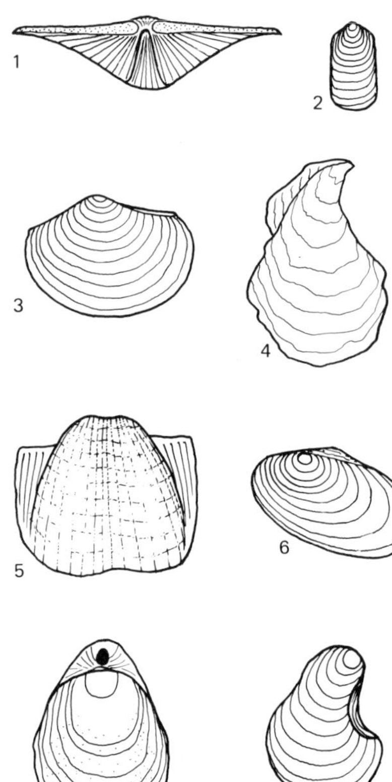

Fig. 83
Some fossil shells

Fossils are often preserved as moulds or casts in older rocks. Moulds are common in shales. They are made by the shell as it is buried and compressed in muddy sediments. Percolating water dissolves away the calcium carbonate but doesn't affect the mould.

In the activity, you made a cast from the mould by filling the space left by the shell with plaster. In nature, the space might be filled with another mineral such as silica (silicon dioxide).

You now know one reason why fossil shells aren't made from the calcium carbonate of the original shells. Another reason is that the calcium carbonate can be replaced crystal by crystal by a different mineral, instead of the whole shell being first dissolved away to leave a space.

This kind of fossil, along with the ones made from the original calcium carbonate, are more useful than moulds and casts because the markings on the shells are better preserved.

Recognizing fossils

You will see from Fig. 77 that fossils could be divided into ones formed by animals and ones formed by plants. The animal kingdom can be further divided into animals with backbones (vertebrates) and animals without backbones (invertebrates). There are far more invertebrate fossils than vertebrate fossils, partly because many vertebrates lived on land.

Some invertebrate fossils are shown in Fig. 83. They all look something like modern animals with two shells that you can find on the seashore. But half of them belong to a group of invertebrates called brachiopods, which are now almost extinct. The other half are lamellibranchs ('clams') and large numbers of them cover some of our beaches today.

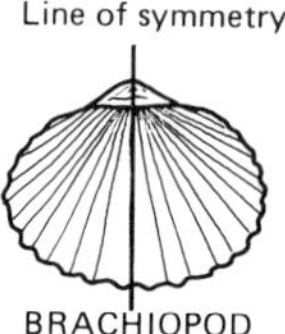

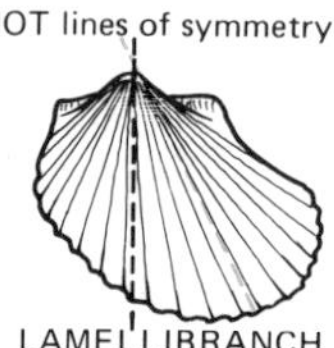

Fig. 84
One difference between a brachiopod and a lamellibranch

One way of telling brachiopods from lamellibranchs is by looking for a line of symmetry down the shell. For brachiopods, an imaginary line can be drawn down a single shell which divides it into two identical halves (Fig. 84). There is no line of symmetry down a single lamellibranch shell.

- Sort out the numbered shells in Fig. 83 into a group of brachiopods and a group of lamellibranchs.
- Is the shell in Fig. 85 (very like the one used to advertise a brand of petrol) a brachiopod or a lamellibranch? (Look at the diagram very carefully before you answer this question.)

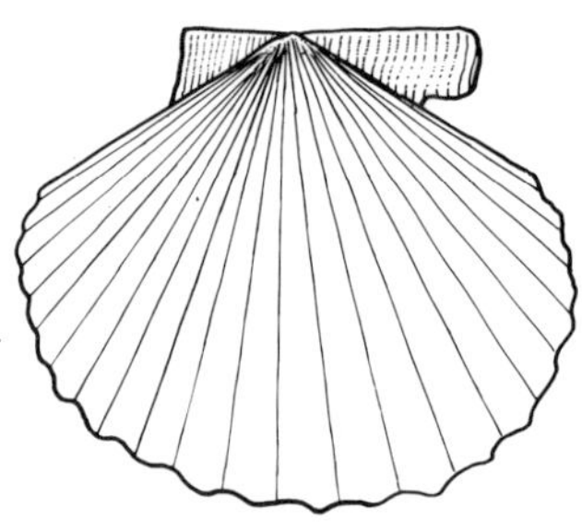

Fig. 85

The key in Fig. 86 will help you to decide how to classify any of the common invertebrate fossils you might find. The first feature to look for is the shape of the shell or skeleton.

Tribolites and graptolites are put into a separate diagram. They are quite easy to recognize. Trilobites have skeletons which are segmented (divided into several parts), and graptolites look like tiny saw blades.

Activity 14 Using keys to identify some fossils

Follow the key (Fig. 86) with your fossil until you reach a diagram.

- Does each of your fossils look quite like the diagram?

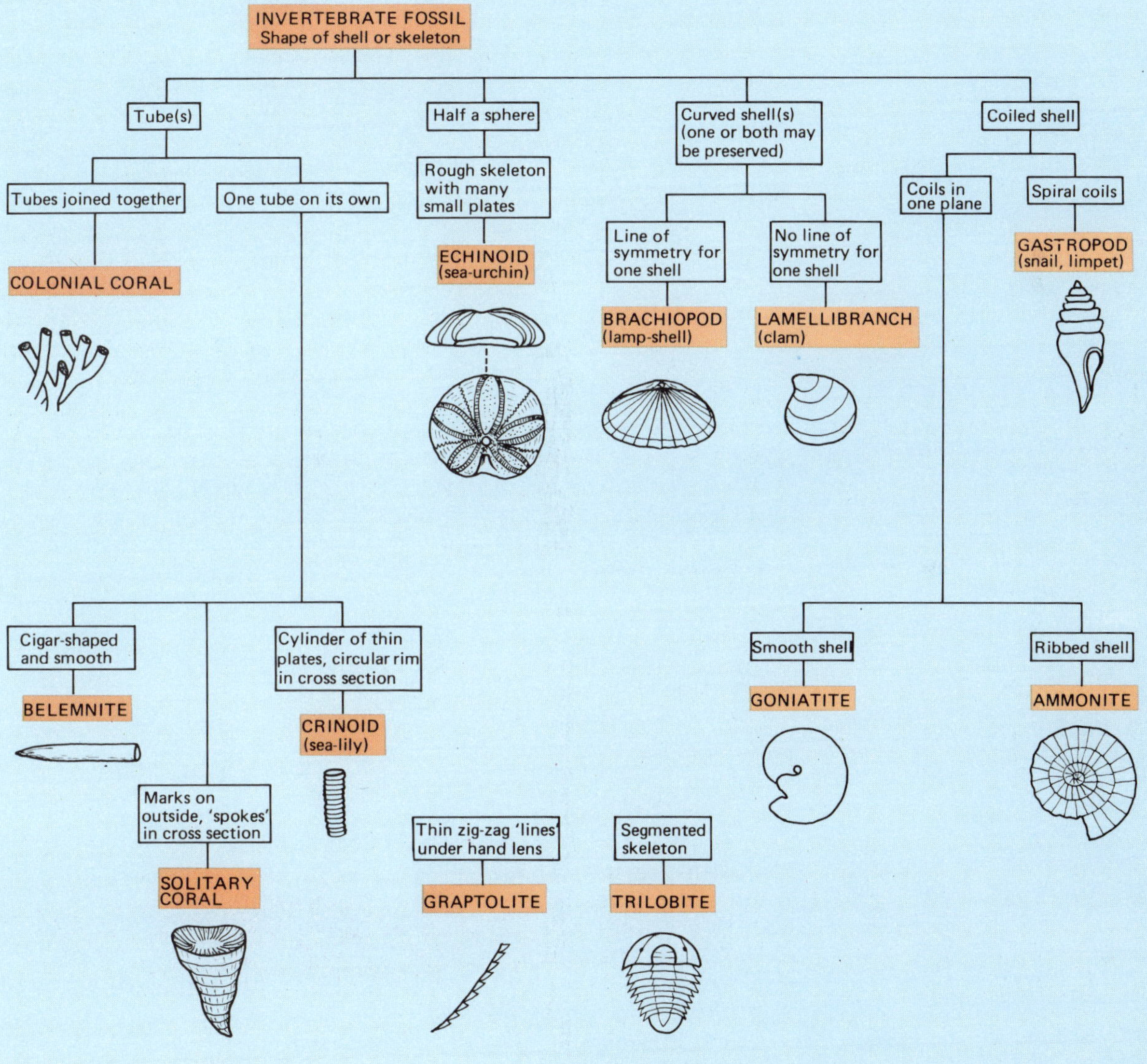

INVERTEBRATE FOSSIL
Shape of shell or skeleton

- Tube(s)
  - Tubes joined together
    - COLONIAL CORAL
  - One tube on its own
    - Cigar-shaped and smooth
      - BELEMNITE
    - Marks on outside, 'spokes' in cross section
      - SOLITARY CORAL
    - Cylinder of thin plates, circular rim in cross section
      - CRINOID (sea-lily)
- Half a sphere
  - Rough skeleton with many small plates
    - ECHINOID (sea-urchin)
- Curved shell(s) (one or both may be preserved)
  - Line of symmetry for one shell
    - BRACHIOPOD (lamp-shell)
  - No line of symmetry for one shell
    - LAMELLIBRANCH (clam)
- Coiled shell
  - Coils in one plane
    - Smooth shell
      - GONIATITE
    - Ribbed shell
      - AMMONITE
  - Spiral coils
    - GASTROPOD (snail, limpet)
- Thin zig-zag 'lines' under hand lens
  - GRAPTOLITE
- Segmented skeleton
  - TRILOBITE

Fig. 86 A key for invertebrate fossils

The keys are only a rough way of sorting out fossils into groups. Each of the groups could be divided up further: there are many different kinds of trilobites, many different kinds of lamellibranchs and so on. For this kind of sorting out, you need to look very closely at the markings on the shells or skeletons.

Activity 15 Sorting out some lamellibranch shells

Use a selection of about six fossil lamellibranch shells. Give each one of them a label A, B, C and so on.

First, look at the shells (using a hand lens if necessary), and write down a list of different features that you notice. Fig. 87 may help you to do this.

Now choose one of these features and sort out the shells into two groups. Keep on using other features until you have finally found ways of telling each shell from all the others.

The key in Fig. 88 shows how five lamellibranchs P, Q, R, S and T might be sorted out.

- Draw a key of this kind for your own results.

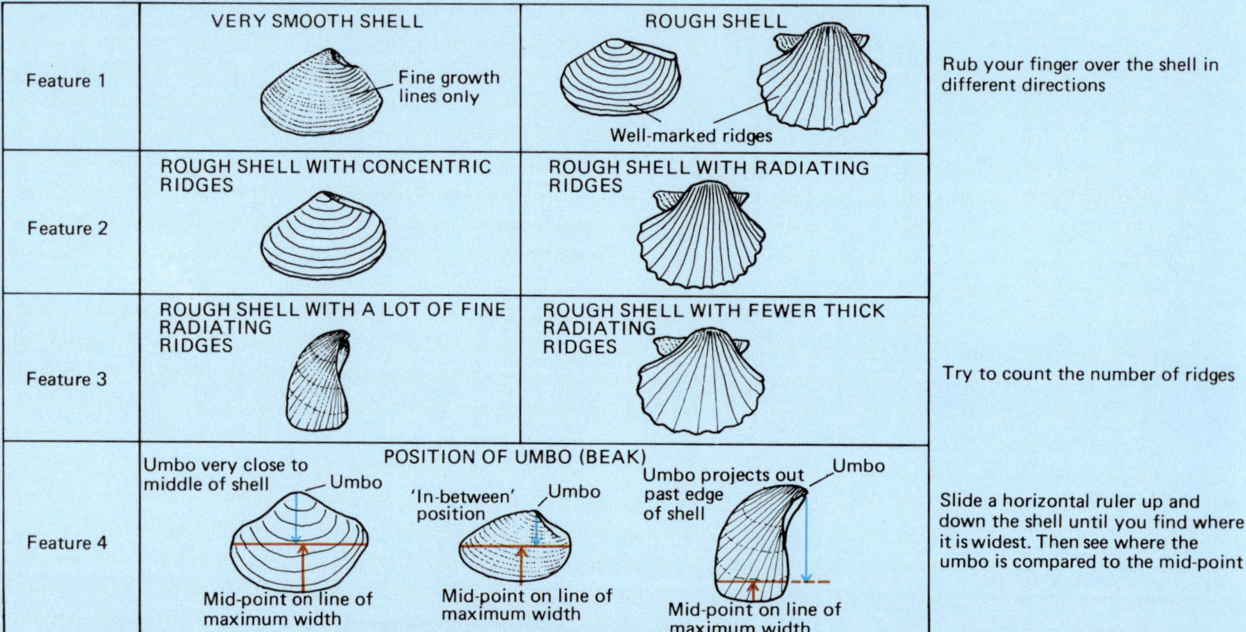

Fig. 87 Some identifying features for lamellibranch shells

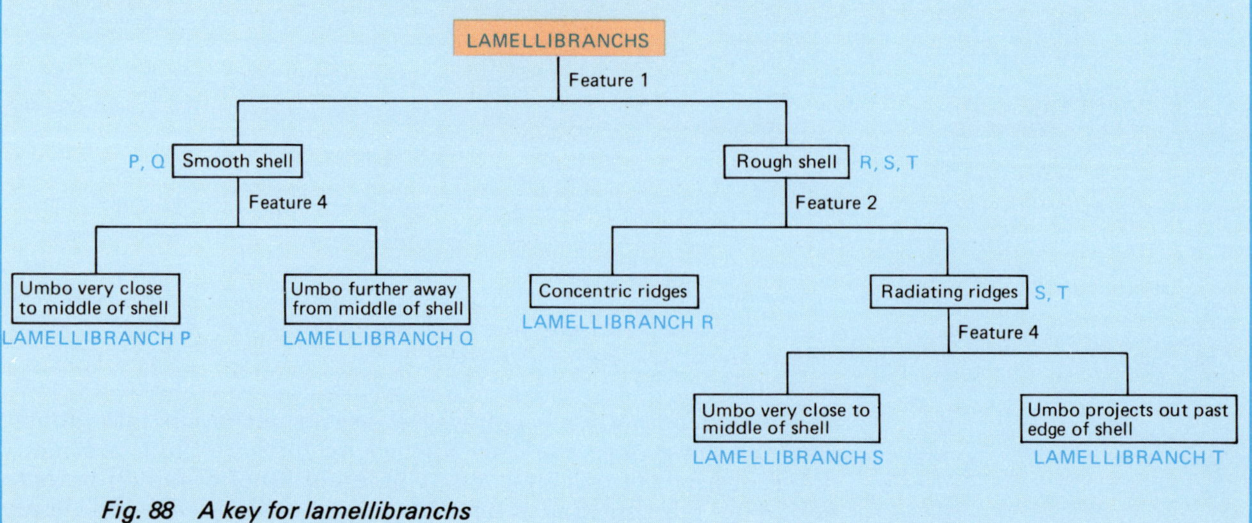

Fig. 88 A key for lamellibranchs

Fig. 89

Each separate shell is a different species of lamellibranch. And each species has a Latin name, not always very easy to remember (Fig. 89)! If you want to know the name of any fossil you collect, you will have to try and match your specimen with the diagrams in a reference book of fossils.

You might like to do Activity 15 again, this time using a selection of other fossils such as ammonites.

Fossils and time

Fossils can be used to estimate the ages of rocks. This was first discovered by an Englishman called William Smith in the early 1800s. He was an engineer and travelled about the country planning where canals and bridges should be built. But he soon got interested in the sedimentary rocks and the fossils they contain.

At that time, a lot of geologists believed that rocks and fossils were all formed during Noah's flood, which would suggest that all sediments and remains of living forms would have been jumbled up. So the idea of rocks in the same exposure having different ages was very new, as was the idea that there was any sensible pattern of events to be read in the rocks.

William Smith collected fossils all over Britain and tried to match those from one exposure with those from other exposures. Layers of sedimentary rocks are usually in order of their age, with the oldest layer at the bottom and the youngest at the top. This means that fossils found at the top of his exposures must be younger than the ones found at the bottom.

He assumed that two rocks in different places must be of the same age if they contained the same fossils. This must be so, even if the rocks themselves are very different—it doesn't matter that one might be a sandstone and the other a shale.

The matching of fossils in different places is called correlation. Through correlation, William Smith was able to make a list of fossils in order of their age. He called this list a 'geologic column'.

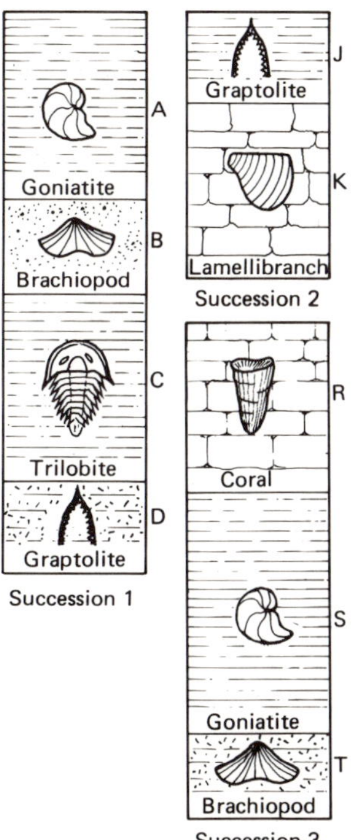

Fig. 90
Three successions of rocks with fossils

You can try to build up a geological column for yourself. Fig. 90 shows three successions of rocks and their fossils in three different places.

- Write down the first succession of rocks in your notebook using the letters from the diagram underneath one another. Put the other successions by the side of this.
- Draw lines between the letters to match up rocks of the same age in all three successions.
- Make a list of all the rocks and fossils in the order that they were probably formed, putting the oldest one at the bottom. This is your geological column.

It isn't really as easy as this to construct a geological column. For one thing, rock successions are sometimes the wrong way up. It is important to know how to recognize these. In Fig. 91, the geologist has noticed that each bed gets coarser grained towards the top.

- Why must this prove that the rock succession is the wrong way up (see page 5)?

Fig. 91
Using graded bedding to show that a rock succession is the wrong way up

Another reason is that some rocks contain different fossils even though they are of the same age. So you find fossil sea shells in rocks deposited under the sea but not in rocks formed on land. Instead, rocks formed on land are likely to contain the bones of land animals or the remains of plants.

But, even so, careful work by people like William Smith has allowed us to divide up geological time into periods. Today we can use radioactive dating (page 14) to put actual dates in millions of years to these periods.

Fig. 92 is the geological time-scale put alongside the life-span of some of the invertebrate groups. Each period has a name. The thickness of the block for each fossil group gives you some idea of how common this kind of life was at the time.

Fig. 92
Life through geological time

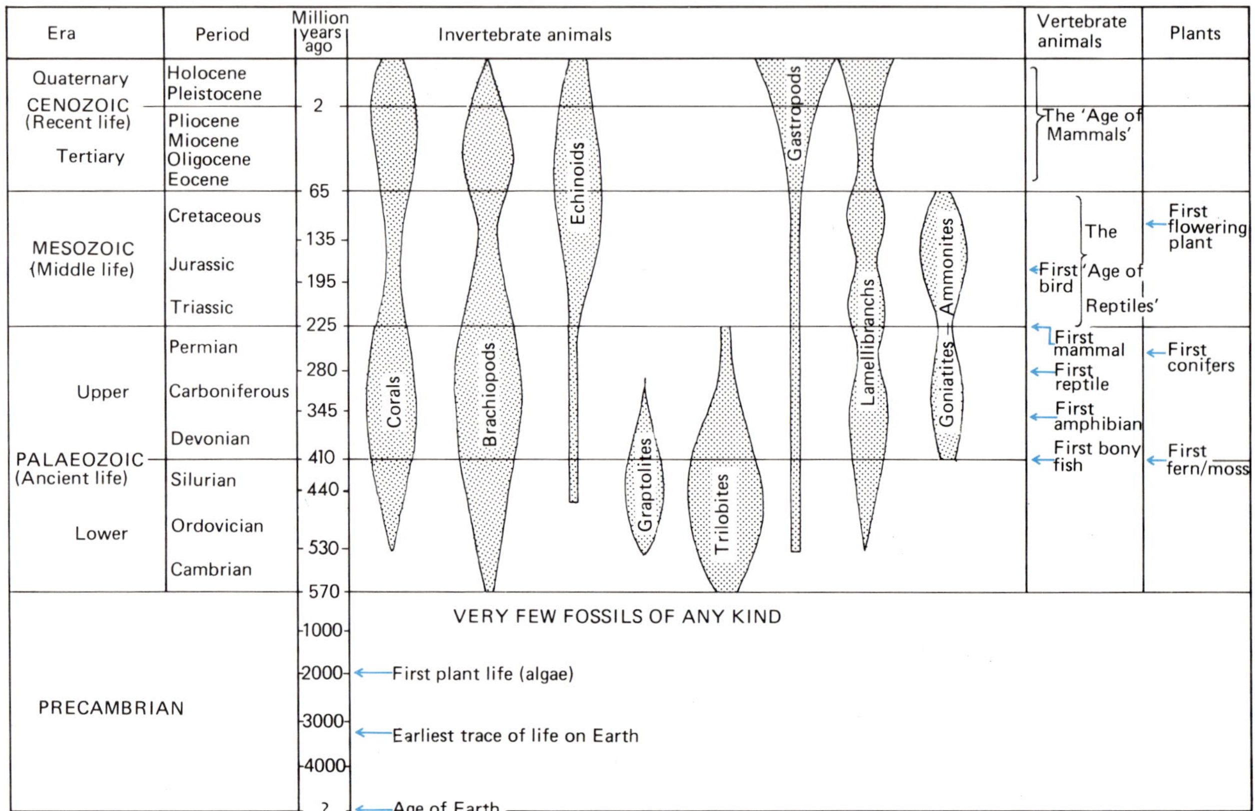

Notice that the Precambrian is much longer than all the other periods put together. This is because there are so few fossils in these rocks that it is difficult to divide it up any further.

The periods are grouped into longer units of time called eras.

- Make a list of the most common invertebrate animals in
  - (i) the Palaeozoic era
  - (ii) the Mesozoic era.
- Underline the names of the groups in the first list that became extinct before the Mesozoic era.
- What were the most common vertebrate animals in
  - (i) the Mesozoic era
  - (ii) the Cenozoic era?

Everybody finds it difficult to imagine the enormous length of geological time. Even one million years seems to us to be a very long time indeed. And yet this is only about 1/4000th of the total age of the earth. When a geologist talks about 'recent events', he may mean something that happened 10 million years ago!

An exercise may help you with this problem. Suppose that you condense the age of the earth into one year of 365 days. To make the maths easier, take the real age of the earth to be 3650 million years.

One day now represents 10 million years.

- How many 'days' have there been since the beginning of the Cambrian?
- About how many 'days' did the Precambrian last?

Man has been living on the earth for only a short length of geological time. In Britain, the earliest man probably dates from about half a million years ago.

- How many 'hours' in a 'day' does this represent?

Perhaps you may live to be 100. There are about 100 000 seconds in a day.

- How long would you have lived on the condensed time-scale?

Fossils and past climates

Fossils aren't only clues about what life used to be like in the past. Sometimes they tell us something about past climates as well. This is true of the fossil pollen grains from trees and other plants.

Different trees grow in different climates (Fig. 93). And each kind of tree has its own particular kind of pollen grain that can be recognized under a microscope (Fig. 94).

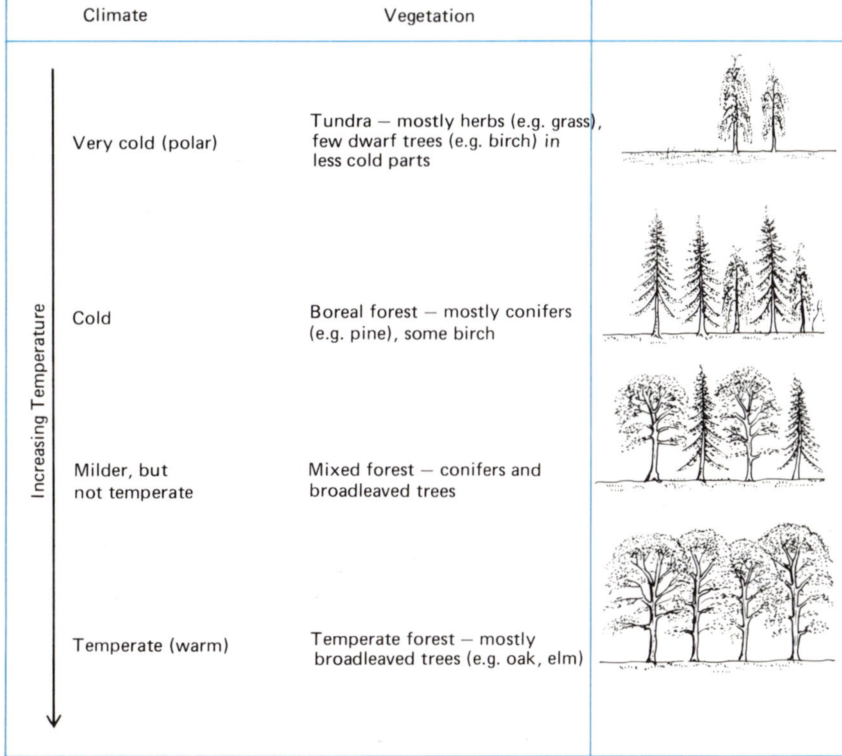

Fig. 93
The vegetation in polar to temperate climates

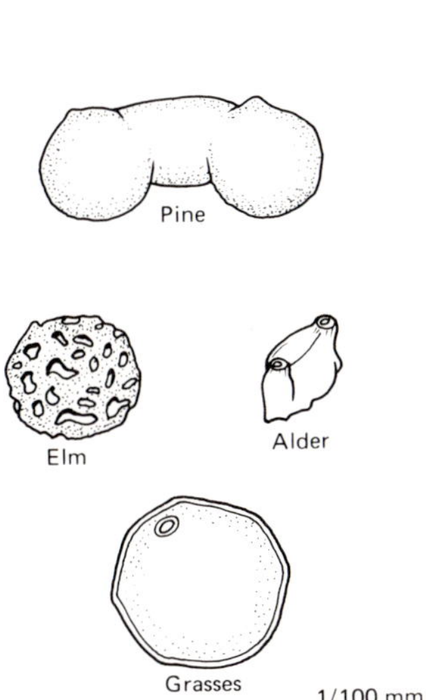

Fig. 94
Some kinds of pollen grains

Various parts of trees and plants have been preserved as fossils. But you may wonder how it comes about that something as small as a pollen grain can also be found in rocks.

Pollen comes from the flowers of a plant. Unlike the rest of a flower, it is made of a material that doesn't easily rot away. It is dispersed by the wind and gets deposited in sediments under water or in peat bogs.

Geologists have to separate the fossil pollen from the rocks or the peat, and then make a count of each kind of grain. If a pollen count is made at various levels in the deposit, you get a record of how the climate has changed during the time the deposit was being formed.

Here is some information about the pollen counts for deposits in south-east England which are between 128 000 years and 115 000 years old (Fig. 95). The thickness of the block for each kind of grain shows how common its parent tree was at the time.

- Can you describe the very big changes in climate that must have taken place over those 13 000 years? Use Fig. 93 to help you.

There are some gravels under Trafalgar Square in London about 120 000 years old that contain the remains of elephant, hippopotamus and rhinoceros. No such remains are found in any younger deposits.

- Is the pollen evidence in line with the idea of big game roaming over this part of England at that time?

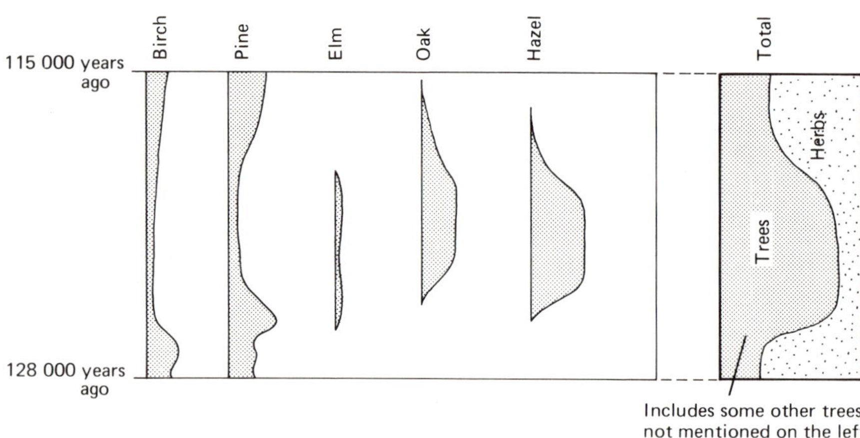

Fig. 95
Pollen counts in some Pleistocene deposits of south- east England

- Why would all these animals have disappeared a few thousand years later?

The world's climate has swung to and fro like a pendulum between very cold conditions and warmer conditions over the last 2 million years or so. We call this period the Pleistocene (Fig. 92) or the Ice Age.

Fossil pollen shows that the British climate began to warm up again about 10 000 years ago. About 6000 years ago, even the Lake District mountains were covered by oak and elm forests up to a height of 600 metres. But from that time, the climatic pendulum seems to have swung back a little towards cooler conditions.

Going much further back in time, there is plenty of evidence that the British climate has changed even more dramatically than in the last 10 000 years. You can read more about these changes in Further study 4: *Past environments.*

Stories in the rocks

Much of the study of rocks takes place out of doors (in the field, as geologists say). This section is a series of real fieldwork problems from different parts of the country. No doubt you will meet others when you go into the 'field' yourself.

As you are thinking about these problems, you will be unravelling the piece of earth's history that is locked up in the rocks of each area.

Fig. 96
A headland of breccia at Thurlestone, south Devon

Ancient beach or ancient scree?

Fig. 96 shows a small headland of red-coloured rock on the coast of South Devon. There isn't much of this rock around here and the only other exposure is in a sea stack in the bay. The rock is made from a large number of pebbles with grains of material in between (Fig. 97). It is an example of a breccia (page 35).

What is the story of this rock? To answer this question, you need to know where the pebbles in it have come from and how they got there.

There are three main kinds of pebbles in the rock—schist, quartz and shale. The first two kinds are quite large pebbles with some sharp edges as well as some more rounded ones. The shale pebbles are smaller but still have some sharp edges.

Fig. 97
A close-up of the breccia

These sharp edges are an important clue. Think about the difference between pebbles on a beach or in a river and pebbles in a scree on the slopes of a mountain.

- Which pebbles have no sharp edges and are very rounded?
- Which pebbles have sharp edges and aren't rounded at all?
- Is the breccia more likely to be an ancient beach or river deposit, or an ancient scree deposit?

Look at the map showing the rocks in this part of south Devon (Fig. 98). It isn't obvious from this where the quartz pebbles may have come from, but you should see that the schist and shale pebbles could have come from rocks close to where the breccia has been deposited.

- Why do we know that the breccia must have been formed later than the Devonian period?

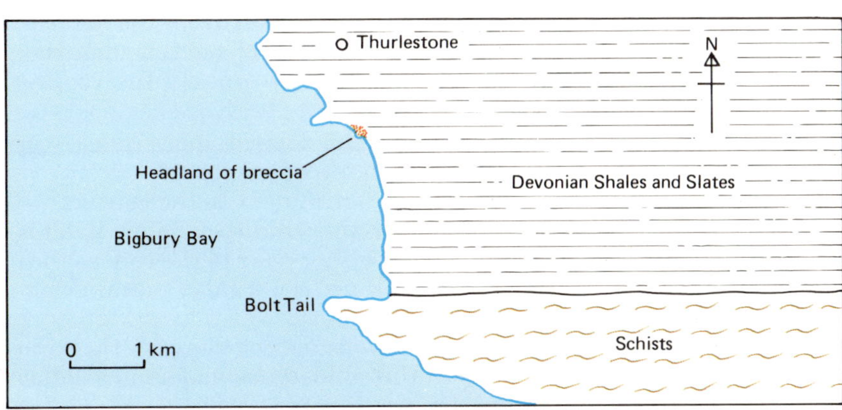

Fig. 98
A simplified geological map of part of south Devon

If the breccia is a kind of ancient scree deposit, the south Devon must have looked very different at that time. Instead of a fairly low-lying rocky coast, there must have been high mountains of schist to the south.

In fact, this picture may not be quite complete. The pebbles in a scree deposit have no rounded edges because they haven't moved very far at all.

But the pebbles in the breccia are partly rounded. This may mean that the scree was moved over a short distance, perhaps by a sudden torrent of water flowing down the mountain. So some edges of the pebbles were rounded a little during the brief journey, though many edges still stayed sharp.

An igneous problem (or The Great Whin . . . What?)

Quite a lot of northern England is made from limestones, sandstones and shales of Lower Carboniferous age. But if you take the tourist track along the deep valley of High Cup Nick, near Appleby in Cumbria, you meet an impressive semicircle of crags at its head which are made from a completely different rock (Fig. 99).

Fig. 99
High Cup Nick

Fig. 100 is a close-up photo of this rock.

- What differences can you see between this rock and a sedimentary rock? (Think about the presence or absence of well-marked bedding planes and joints.)

In fact, the rock is made from crystals and is igneous, not sedimentary. It is dolerite, a rock quite like basalt except that its crystals are just big enough to be seen with the help of a hand lens.

In one way, this dolerite at High Cup Nick is very like a sedimentary rock. It is just one more layer in a succession of layers of limestones, sandstones and shales. All these layers dip gently to the north-east. The dolerite is said to be conformable with the other rocks.

The lower contact of the dolerite with some sandstones and shales is shown in Fig. 101. The rocks look much the same in colour, but you should now know how to tell them apart. If you could look at samples of both kinds of rock close to the contact, you would find out two more things about them:

1. the crystals of the dolerite are small next to the contact and larger a few metres away from it (Fig. 102)
2. the sandstones and shales are much tougher (more difficult to break up with a hammer) than rocks further away from the dolerite

All igneous rocks are formed by the cooling of molten material, and the size of their crystals depends on how quickly the cooling takes place.

- Did the molten material cool to form the dolerite more quickly or more slowly at the contact than further away from it?
- Why should the rate of cooling be different next to the contact?

Also, the hot molten material 'baked' the solid rock below it. This is why the sandstones and shales are tougher near to the contact but stay unchanged further away. The 'baking' is a kind of metamorphism (page 24).

- Does what you know about the lower contact suggest that the dolerite is older or younger than the sandstones and shales below it?

Fig. 100
An exposure of dolerite near High Cup Nick

Fig. 101
A close-up of the lower contact between the dolerite and the sedimentary rocks

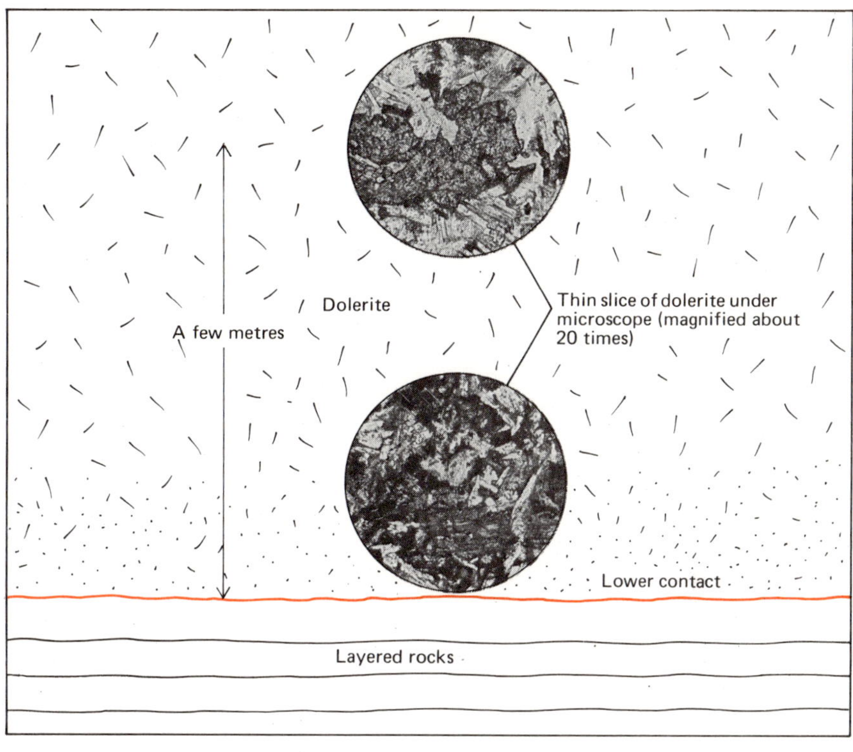

Fig. 102 The variation of crystal size in the dolerite close to the lower contact

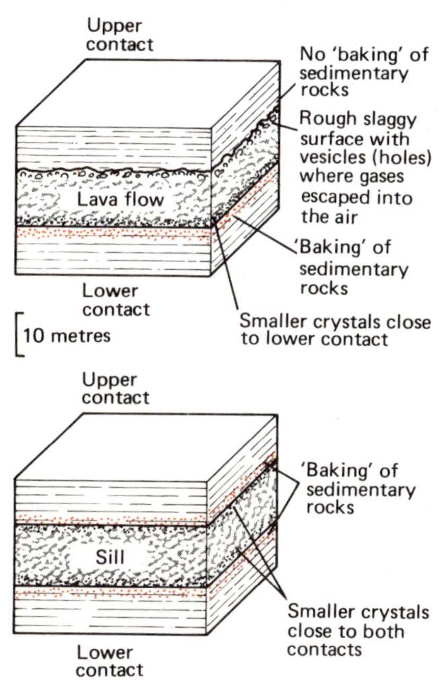

Fig. 103
Some differences between an ancient lava flow and a sill

This is the same answer you would give if there were a layer of sedimentary rock on top of the sandstones and shales.

An igneous rock that is conformable with the sedimentary rocks around it can be one of two things (Fig. 103)—a lava flow (like the Giant's Causeway in Fig. 29) or a sill. Some of the differences between a lava flow and a sill can be seen at the upper contact.

- Use Fig. 103 to decide what these are. Make a list of them.

The differences come about because of the different ways in which a lava flow and a sill are formed.

A lava flow comes from a volcano and spreads out across the surface of the land (or the sea floor). Though there are rocks below a lava flow, its top surface is open to the air. It is here that the gas bubbles collect and get trapped during cooling. This makes the top of the rock rough and slaggy. The next layers may be more lava flows or sedimentary rocks. Whatever they are, there can be no 'baking' of them because the hot material has long since cooled down.

- Is the lava flow in Fig. 103 younger or older than the sedimentary rock above it?

A sill is formed by the cooling of molten material that has been injected along the bedding planes of sedimentary rocks but never reaches the surface. There is rock above and below the sill. So what happens at the lower contact also happens at the upper one.

- Is the sill in Fig. 103 younger or older than the sedimentary rock above it?

The dolerite at High Cup Nick is about 30 metres thick and an upper contact can be found just above the line of crags shown in Fig. 99. It turns out to be like the lower one, so proving that the dolerite is a sill. Geologists call it the Great Whin *Sill*—the problem is solved.

The sill is obviously younger than the Lower Carboniferous rocks that it has been injected into. But how much younger is it?

One way to find out is to look for pebbles of the dolerite in younger conglomerates or breccias.

- Would the sill be older or younger than such a conglomerate or breccia?

Only a few kilometres south of High Cup Nick there is a Lower Permian breccia containing some pebbles of rotted dolerite.

- *If* these pebbles were eroded from the Great Whin Sill, what can you say about its age?
- Use the geological time-scale (see back cover) to give the Great Whin Sill a rough age in millions of years.

Another way to find the age of an igneous rock is by radioactive dating. The radiometric age of the Great Whin Sill is known to be 295±6 million years, that is somewhere between 289 and 301 million years.

- How well does this age match with your rough estimate?

The Great Whin Sill is a very good example of a sill. It extends for about 130 kilometres across northern England (Fig. 104) and keeps an average thickness of 30 metres. There are several well-known beauty spots on the way (Figs. 105 and 106), and the dolerite is quarried in some other places for roadstone.

- Why does the sill form scarps and headlands?

Fig. 104
Outcrops of the Great Whin Sill in northern England

Fig. 105
The scarp formed by the Whin Sill in Northumberland. Here the Roman Wall follows its crest

Fig. 106
Dunstanburgh Castle in Northumberland. The castle stands on a small headland of Whin Sill, and rocks in the foreground are also dolerite.

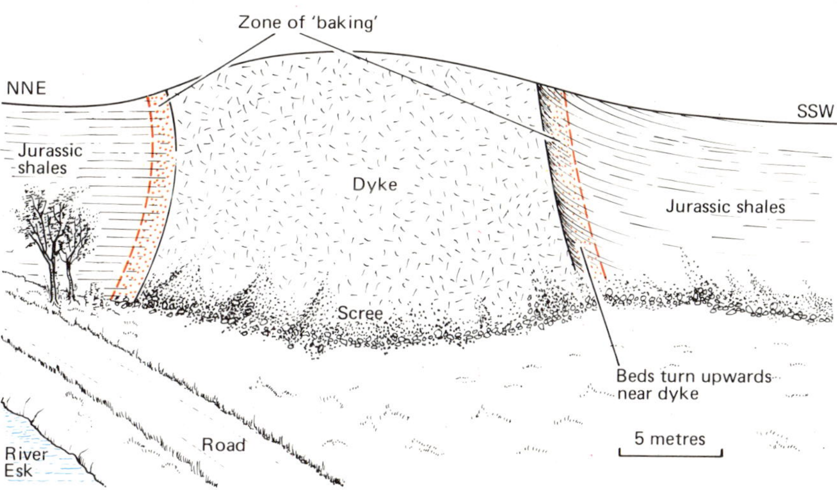

Fig. 107
A quarry in the Cleveland Dyke near Egton Bridge, North Yorkshire

Large sheets of molten material a few metres thick can move up through vertical cracks as well as along bedding planes. These cool down to form dykes. Unlike sills, dykes are never conformable with the sedimentary rocks in which they are found.

Fig. 107 shows a large dolerite dyke in North Yorkshire called the Cleveland Dyke. You can see that the molten material must have been pushed upwards and across the beds of Jurassic shales.

- How wide is the dyke?
- What has it done to the surrounding shales?
- Is the dyke likely to have small crystals close to one of the contacts or close to both of them?
- Why is it certain that the Cleveland Dyke can't be the same age as the Great Whin Sill?

The radiometric age of the Cleveland Dyke is 58 ± 1 million years. This means that it must have formed during the early part of the Cenozoic (Tertiary) era.

Fig. 108
Arcow quarry, Ribblesdale, North Yorkshire

Time-gaps

Look at the quarry face in the photo (Fig. 108). There is a white limestone on top of a grey-green sandstone. The break between these rocks is a grassy ledge about two-thirds the way up the face from the bottom. The layers of limestone look horizontal while the layers of sandstone dip quite steeply.

- Draw a sketch of this face showing the dip of each rock.
- Which is the younger of these rocks?

There are fossils in each of the rocks—corals and brachiopods in the limestone and graptolites in the sandstone. These prove that the limestone dates from the Carboniferous period and the sandstone from the Silurian period (page 44). The Carboniferous period doesn't directly follow the Silurian period.

- Work out from the geological time-scale the size of the time-gap between the Carboniferous limestone and the Silurian sandstone.

The story behind the rocks in the quarry face links the sudden change in the dip to the large time-gap. Geologists call the kind of structure in Fig. 108 an unconformity. The next activity should help you to decide how the unconformity was formed.

Activity 16 Making a plasticine model of an unconformity

Follow the instructions in Fig. 109. Put a piece of thin paper between each layer of plasticine.

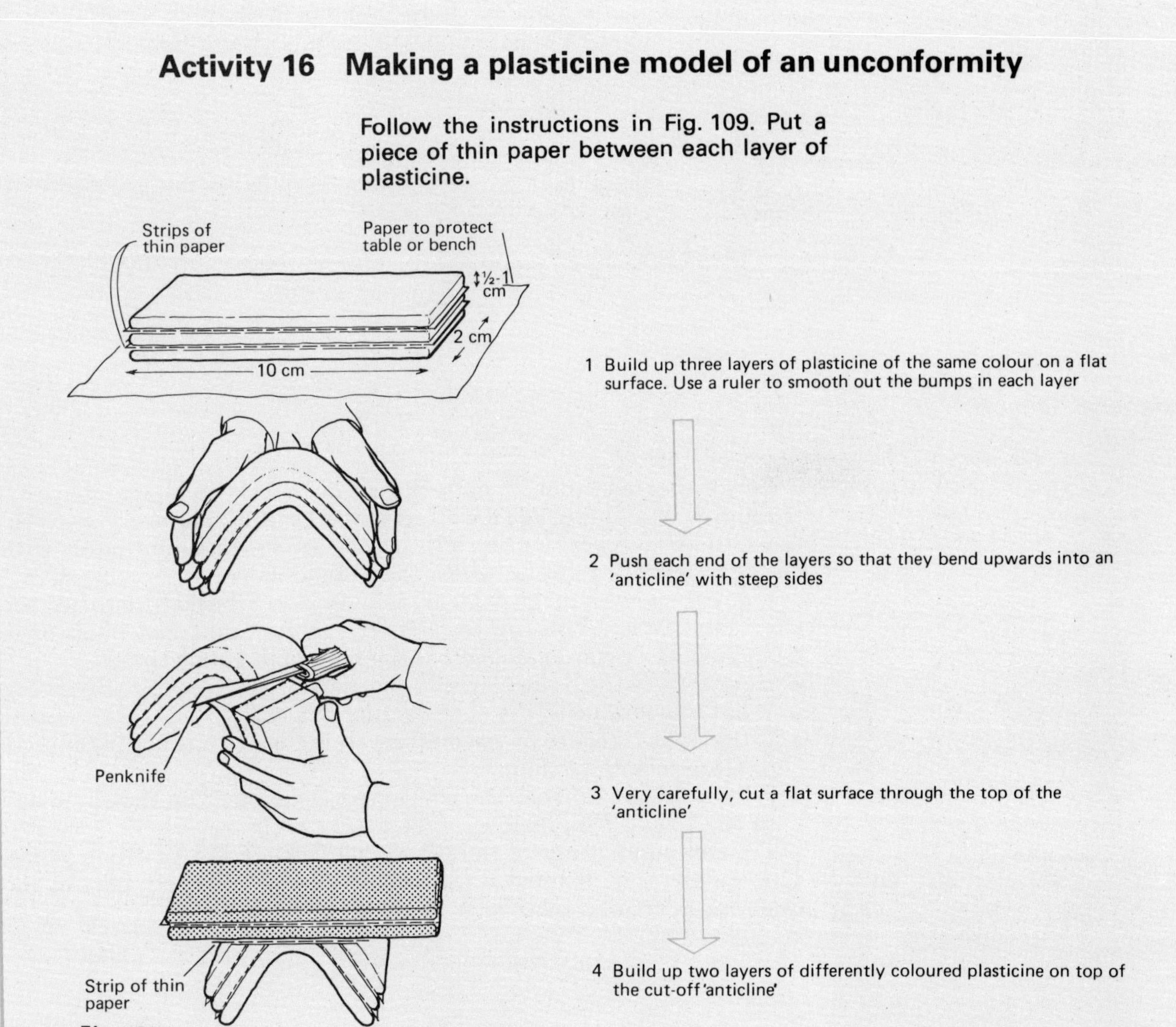

Fig. 109

Imagine that the layers of plasticine in stage 1 are layers of the Silurian sandstone and the layers of plasticine in stage 4 are layers of the Carboniferous limestone.

- What process must have made the layers of the Silurian sandstone tilt at a high angle?
- What real-life process is like the cutting of the lower layers in stage 3?

This process is what must have been happening during the time-gap between the Carboniferous and Silurian periods.

- What process must have formed the flat layers of the Carboniferous limestone on top of the tilted Silurian sandstone?
- Use this activity to tell the story of the rocks in the quarry face. Begin with the forming of the Silurian sandstone and end with the Carboniferous limestone.

An unconformity is formed when there is a break between the deposition of one set of rocks and the deposition of another much younger set. During this break, other processes in the rock cycle (page 26) are taking place. Mountain-building lifts the older rocks out of the sea while erosion works against this by slowly lowering the rocks down to sea level again. So the unconformity in the rocks represents the time when the area was land, not sea (see Further study 5: *Rock cycles in the past*).

The quarry face in Fig. 108 is in the Pennines near Horton in Ribblesdale. The map (Fig. 110) shows some of the exposures of Carboniferous rocks and older rocks (of Precambrian, Ordovician and Silurian age) that can be found in the area.

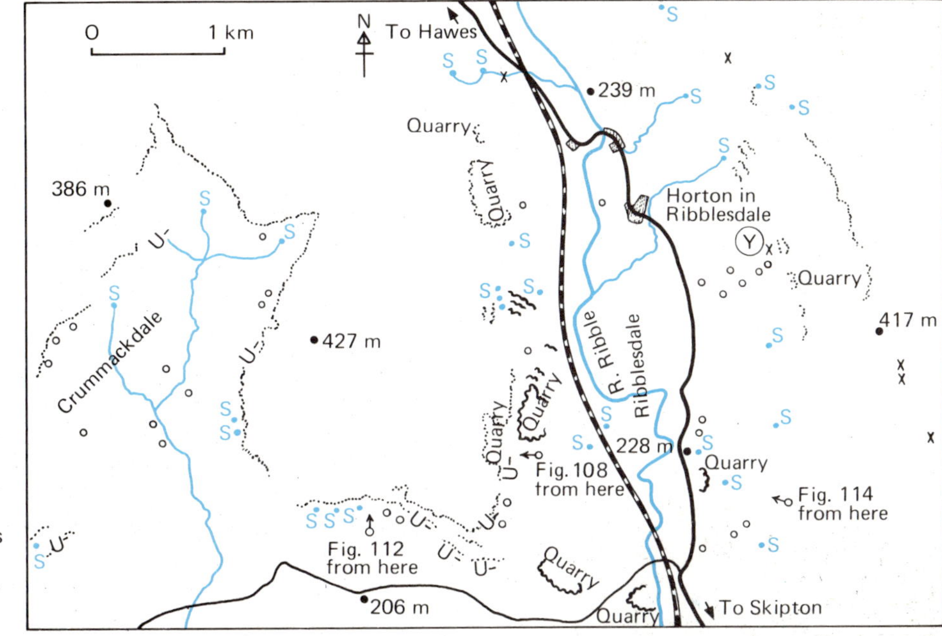

Crag
Other exposure } Limestone

Crag
Other exposure } Older rocks

U- Exposure of unconformity

S Spring line

Fig. 110
A sketch map showing some of the rock exposures in the Horton in Ribblesdale area

The unconformity itself is only exposed in a few places. In others, you have to guess where it is by finding nearby exposures of the Carboniferous limestone and the older rocks. You can then draw a line somewhere between the two (Fig. 111).

Fig. 111
The line of the unconformity.

a) Carboniferous limestone
b) Line of unconformity
c) Older rocks
d) Marshy ground

Fig. 112
Marshy ground near the line of unconformity. Look at Fig. 110 to find the location of this photo

Sometimes it also helps to look for spring lines—places where the ground is marshy because a stream is starting (Fig. 112). This could be formed because water in the limestone can't carry on moving through the rock. It is stopped by the much less permeable older rocks below the unconformity. But you can't rely on all spring lines in the area marking the unconformity, and so you have to be careful in using this evidence.

At point Y, the unconformity itself can't be seen but there is an exposure of a conglomerate (Fig. 113). This has rounded pebbles of the Silurian sandstone embedded in the limestone. It is certain that if you dug into the ground here you would come across the unconformity. The rock just above an unconformity often has pebbles in it which were eroded from the old land surface as the sea moved across it. So the conglomerate is the remains of an ancient pebbly beach.

You can now follow the line of the unconformity through the area shown on the map.

- On a trace of Fig. 110, mark the frame of the map, the rivers and the crags. Draw in the line of unconformity as accurately as you can. Use a broken line where the position is only a rough one because there are few exposures.

Fig. 114 is a view of the west side of Ribblesdale showing the scarps and quarries of the Carboniferous limestone near to the top of the hill, and the quarries of older rocks lower down.

- What clue is there that the quarry on the extreme right of the photo is in limestone? What is limestone used for?
- What do you think the very tough older rocks like the Silurian sandstone might be used for?
- Try to follow the line of the unconformity across the picture, and then match its position with your line on the map.

You will see from your map that the Carboniferous limestone is found close to the tops of the hills while the older rocks are found in the valleys (Ribblesdale and Crummackdale).

- How has this come about? (If you can't answer this question straight away, cut a valley in your plasticine model from Activity 16.)

a) Carboniferous limestone
b) Conglomerate

Fig. 113
The conglomerate just above the unconformity

Fig. 114
An overall view of the west side of Ribblesdale showing the line of unconformity (see Fig. 110)

This completes the story of the unconformity by explaining the present-day landscape.

Now think again about your plasticine model. You made the 'older' layers with the same colour of plasticine. Suppose, instead, that you added a second set of layers with a different colour on top of the other layers; and you then followed the instructions in stages 2, 3 and 4 of Fig. 109 again.

You would get a pattern like the one in Fig. 115.

- What can you say about the layers that lie just below the 'unconformity'?

A detailed study of the Horton in Ribblesdale area shows that the Carboniferous limestone rests on rocks of different ages at different places. For example, it rests on the Silurian sandstone in the quarry face (Fig. 108) but on Precambrian slates just to the north of the village.

This comes about because the older rocks have steeply dipping layers. The structure is called an unconformity with overstep—the younger rock 'steps over' from one older rock to another.

There is another unconformity with overstep shown in the simplified geological map of part of Salop (Fig. 116). The Cambrian, Ordovician and Silurian rocks all dip steeply to the west while the Upper Carboniferous rocks dip gently to the north.

- Make a trace of this map and mark on it the line of the unconformity.
- Which rocks are above the unconformity and which ones are below it?

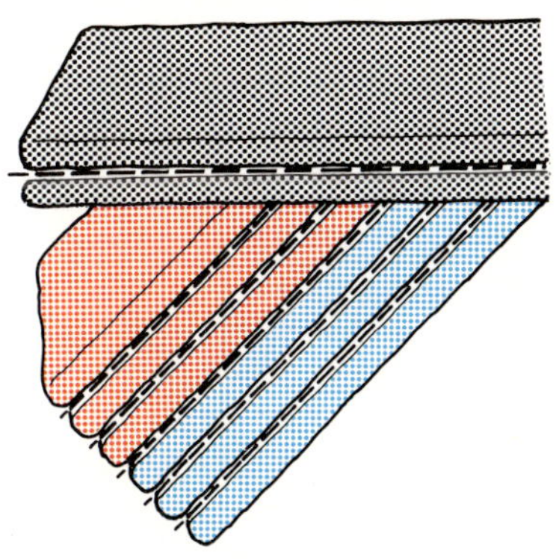

Fig. 115
An unconformity in plasticine

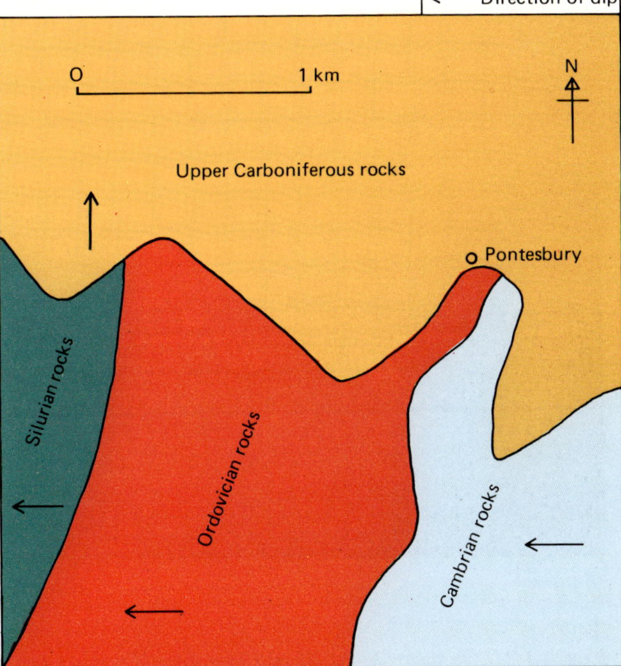

Fig. 116
A simplified geological map of part of Salop showing an unconformity with overstep

Folded rocks

From Newlands Corner near Guildford in Surrey there is a view over the wooded countryside of the Weald (Figs. 117 and 118). The story in the rocks under this area can be worked out by taking a walk down the scarp to a lane just beyond the village of Albury. All the rocks are of Cretaceous age.

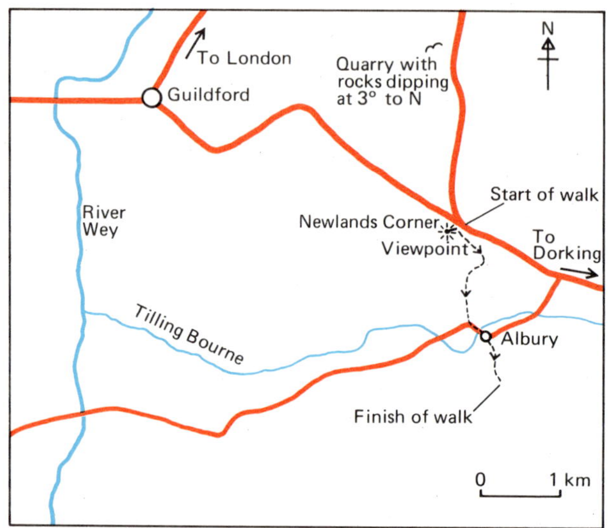

Fig. 117 The Albury area near Guildford, Surrey

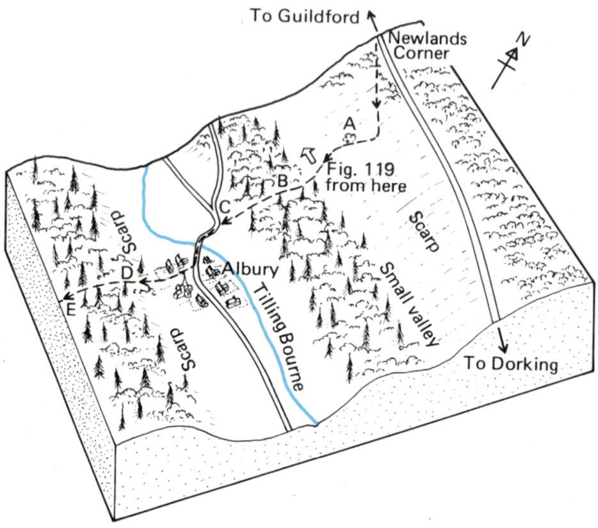

Fig. 118 A block diagram of the countryside around Albury, Surrey (see Fig. 117)

Fig. 119 shows the view on the right during the early part of the walk. Down the first steep slope there is a quarry in white Chalk (A on Fig. 118). After this the land flattens out to make a small valley. The rock under here is Gault Clay.

Now look again at Fig. 119. On the other side of the small valley there is change from fields to heathland with bracken and conifers. Also the clay soil changes to a very sandy one.

- What kind of sedimentary rock forms a very sandy soil?

a) Bracken
b) Sandy soil
c) Conifers
d) Valley in Gault Clay
e) Chalk scarp

Fig. 119 The view west along the small valley below the Chalk scarp (see Fig. 118)

This new rock is exposed in a nearby sand pit (B on Fig. 118). It is very crumbly, brightly coloured and shows some cross-bedding (Fig. 120).

Just before the bottom of the hill, there is another rock in a road cutting (C on Fig. 118). This is a pebbly sandstone which isn't as brightly coloured as the sand in the sand pit. Also it isn't so easily crumbled and it has well-marked bedding planes dipping gently to the north (like the rock in the quarry shown in Fig. 117).

What you now know about the rocks under this area can be put into a section (Fig. 121).

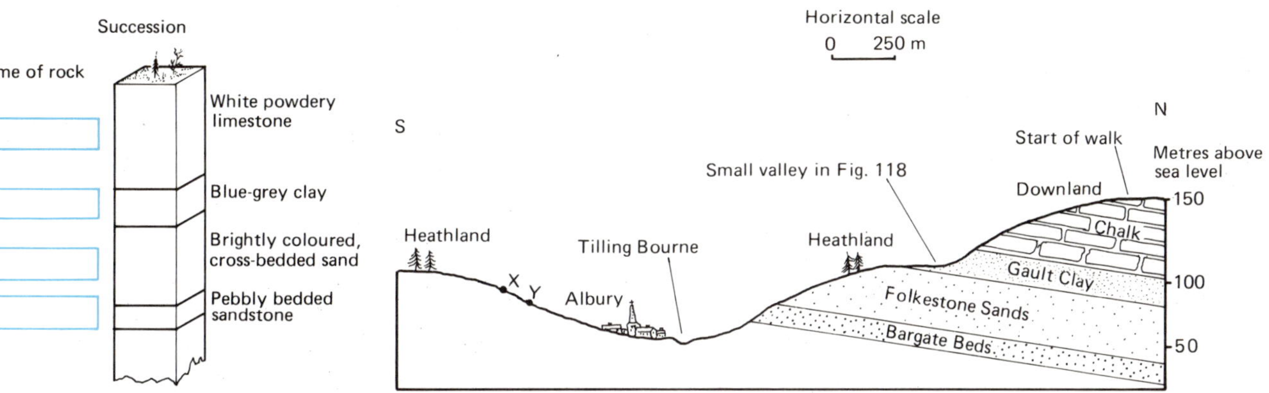

Fig. 121 A geological section through the Albury area along the line of the walk

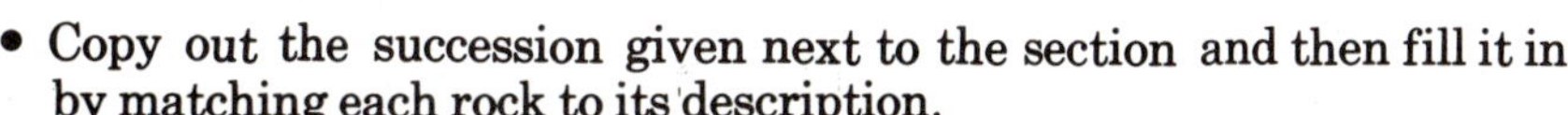

- Copy out the succession given next to the section and then fill it in by matching each rock to its description.
- In the succession, which is (i) the oldest rock and (ii) the youngest rock?
- Suppose that the Bargate Beds dipped gently to the north on the other side of the valley. Use the section to decide where (if anywhere) they would form an outcrop.

There are no more exposures on the walk until you cross the stream and start to walk up the lane on the south side of Albury. Here the pebbly sandstone (the Bargate Beds) can again be seen in the banks of the lane (D on Fig. 118). But this time it is dipping gently to the south.

- Are the Bargate Beds where you expected them to be on this side of the valley?
- What is different about them here?

At the top of the hill, the yellow sand is exposed in the banks of the lane (E on Fig. 118).

- Copy the section (Fig. 121) and show the Bargate Beds forming an outcrop between the points X and Y and dipping gently to the south.
- Mark the Folkestone Sands on top of the Bargate Beds.
- Try to think of a way to explain the change in the direction of dip across the valley.

The rock structure you may draw to answer this last question is called an anticline.

Fig. 120
A sand pit at B in Fig. 118

Activity 17 Making a plasticine model of an anticline

Follow the instructions in Fig. 122. Put a piece of thin paper between each layer of plasticine.

- What is the order of the layers going from the top of the right-hand side of the 'valley' to the top of the left-hand side?

Imagine that the layers of plasticine are layers of different sedimentary rocks.

- Where in the 'valley' can you find (i) the youngest rock and (ii) the oldest rock?

The rock under the middle of the Albury valley is part of the Hythe Beds.

- From your plasticine model, decide where the Hythe Beds come in the succession (Fig. 121).

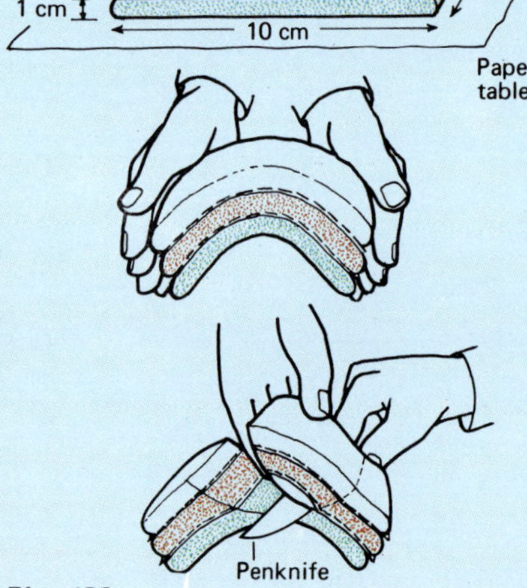

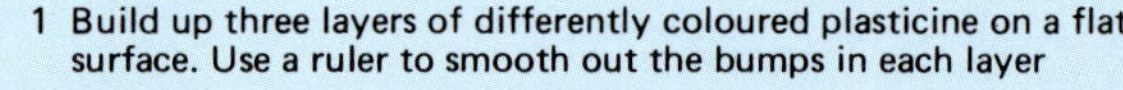

1 Build up three layers of differently coloured plasticine on a flat surface. Use a ruler to smooth out the bumps in each layer

2 Push each end of the layers to that they bend upwards into an 'anticline'

3 Very carefully, cut a 'valley' into the top of the 'anticline'

Fig. 122

You can now think of the anticline as you look back at Fig. 118.

Two scarps face one another across the valley, and the middle of the anticline has been eroded by the stream.

- Use Activity 17 to tell the story of the rocks under Albury. Begin with the deposition of the Hythe Beds and end with the erosion of the middle of the anticline.

Anticlines can be big or small. The one at Albury is quite small but the anticline in Fig. 123 is a lot smaller still.

The Chalk and the Gault Clay in Fig. 118 are on one side of the large Wealden anticline. You may have thought that these rocks must form an outcrop above the Folkestone Sands just to the south of Albury. But they don't. Instead, the Weald is mostly made from rocks like the Folkestone Sands and other, older rocks.

Look at Fig. 121 again.

- What could happen to the direction of dip of the rocks to stop the Chalk and the Gault Clay from forming outcrops just to the south of Albury?

The Chalk and the Gault Clay eventually appear again on the other side of the Wealden anticline about 35 kilometres further south (Fig. 124).

Fig. 123
An anticline in a cutting on the Windermere–Kendal road in Cumbria

The anticline at Albury is a fairly small structure within a much larger one. The rocks in this part of England were folded during the earth movements of Tertiary age that thrust up the great mountain ranges of the Alps and the Pyrenees (see Further study 5: *Rock cycles in the past*).

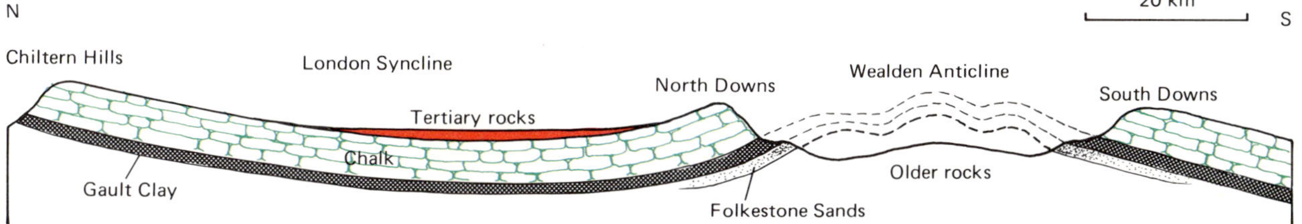

Fig. 124
A geological section through the London syncline and the Wealden anticline

The opposite of an anticline is a syncline like the one in Fig. 125.

- Do you find the oldest or the youngest rocks in the outcrops in the middle of a syncline?

London and the Thames valley are in the middle of a very large syncline which is bounded on both sides by a Chalk scarp (Fig. 124).

Folded rocks are important as places to look for oil, natural gas and water. Oil and gas form in deeply buried rocks and move up towards the surface. Quite often, they get trapped in anticlines higher up the succession.

Fig. 125
The 'Devil's Kitchen' syncline at Cwm Idwal in Snowdonia

Water is different. It comes from above the surface, moves down through the rocks and collects in synclines.

In both cases, the liquid (or gas) only moves through rocks such as sandstones which have plenty of space between the grains. Oil and gas get trapped in the middle of an anticline if the rock above doesn't let them move any further. And water collects in the middle of a syncline if the rock below acts in the same way.

Broken rocks

Scarps, like the one in Fig. 119, can go on without a break for many kilometres. But if you were walking along the crest of the Chalk scarp in North Yorkshire (Fig. 126), you would reach a place where it ends suddenly. The scarp is offset, and it begins again at the same height in another place.

Nearer the sea, there is another offset scarp, this time in a rock called Calcareous Grit. One part of the scarp forms the top of the sea cliff while the other part is further away from the sea.

- Are the Chalk scarp and the Calcareous Grit scarp offset in the same way or a different way?

Why is there such a sudden break in the crests of these two scarps? Imagine a line joining the two offset scarps. This line runs out to sea at a place called Lebberston Cliff. Clues can be found in the rocks of the cliff that help to solve this problem.

Fig. 127 is a field sketch of Lebberston Cliff. You can see that the 'line' appears as a zone of broken rocks, and the rocks are different on either side of the zone.

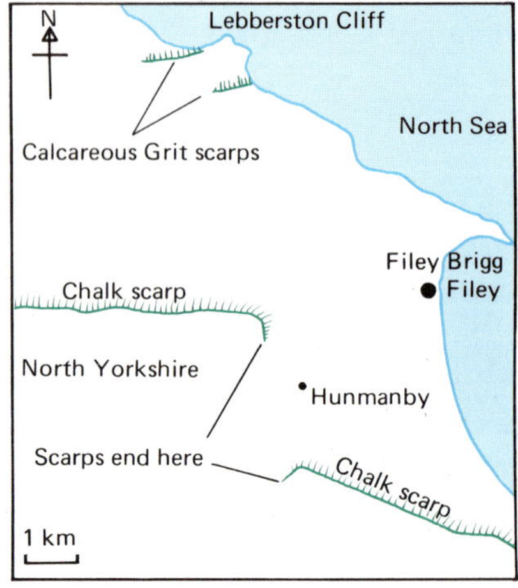

Fig. 126
Offset scarps in North Yorkshire

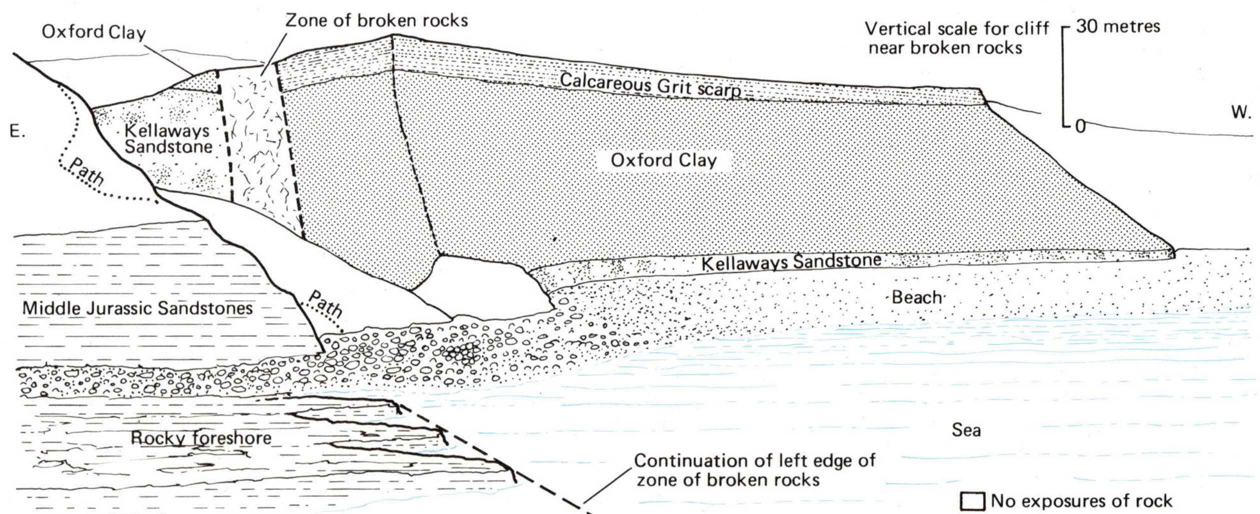

Fig. 127
A sketch of the rocks close to the fault at Lebberston Cliff

- Write down the succession of rocks on (i) the west side of the zone and (ii) the east side of the zone.
- Which rock isn't exposed in the cliff on the west side, and which one isn't exposed on the east side?

The zone of broken rocks in the cliff is called a fault. The rocks on the west side (the downthrow side) have slipped down relative to those on the east side (the upthrow side).

The throw of a fault is the vertical distance that the rocks have slipped away from one another. It can be worked out by matching up the successions on either side of the fault.

- Find the boundary between the Oxford Clay and the Kellaways Sandstone on each side of the fault.
- Use the vertical scale to work out roughly the throw of the fault.

Fig. 128 is a map of the Lebberston Cliff area. It shows the offset scarp of Calcareous Grit and the place on the cliffs where the fault can be seen. The foreshore is rocky on the east side of the fault but is covered with boulders and sand on the west side.

Fig. 128
The Lebberston Cliff area

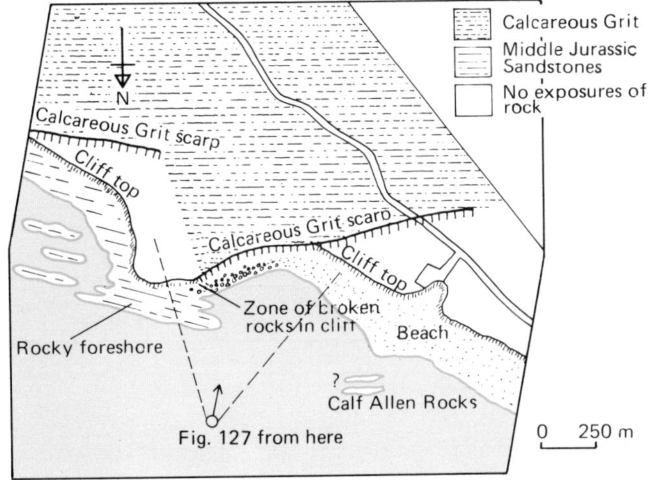

- Make a trace of this map and shade it in the same way.
- Draw the line of the fault on your map and label the upthrow and downthrow sides. Start the fault in the extreme south of the area and end it out at sea.
- Can you think of some reasons why the map doesn't show exposures of Oxford Clay and Kellaways Sandstone?
- Why is the foreshore rocky on the east side of the fault but not on the west side?
- Which rock would you expect to find exposed in the Calf Allen Rocks?

The next activity will help you to find out why a fault can cause a scarp to be offset.

Activity 18 Making a plasticine model of a fault

Follow the instructions in Fig. 129. Put a piece of thin paper between each layer of plasticine.

Imagine that the top layer of plasticine is Chalk or Calcareous Grit. It forms a scarp because the lower layer is a clay and so is more easily weathered and eroded.

- Use Fig. 129 to tell the geological story of how the offset scarps were formed.

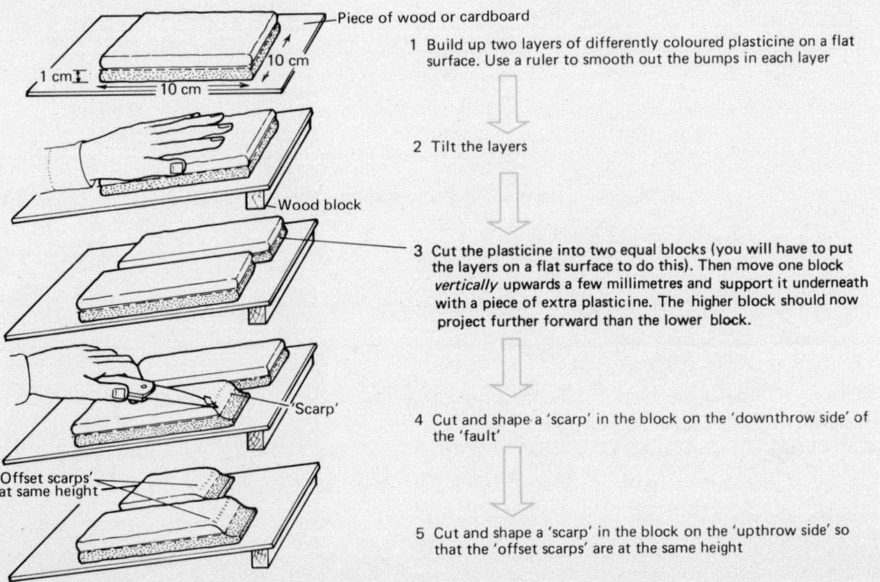

Fig. 129

There is a simplified geological map of part of the North Yorkshire coast in Fig. 130. The Hunmanby fault is shown along with some other faults. Most of them are more or less parallel to the dip but one is nearly at right angles to it.

How can the downthrow side of each fault be found from the map? Look back at the field sketch (Fig. 127) of the Hunmanby fault. At any point along the fault, the rock on the downthrow side is younger than the rock next to it on the upthrow side.

So on a map, you have to use the succession to find the younger rocks. These are always on the downthrow side.

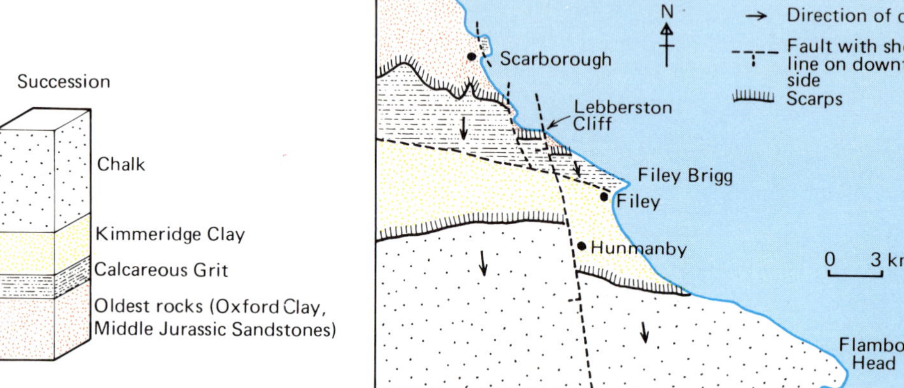

Fig. 130
A simplified geological map of part of North Yorkshire showing some faults

- Make a trace of Fig. 130.
- Use the succession to find the downthrow side of each fault.
- Mark this on the fault using a short line at right angles to the fault line. (This has been done for you on the Hunmanby fault.)

Faults can't always be sorted out quite so easily as at Lebberston Cliff. This is often because the exposures of the fault zone and the rocks on either side of it aren't very good.

There is a fault at Ravenscar, which is on the North Yorkshire coast just north of Lebberston Cliff. Fig. 131 shows that the fault zone at Ravenscar isn't exposed at all. All you can see are some of the rocks on either side. These have been matched up (correlated) using the fossils (especially ammonites) they contain.

Fig. 131
A sketch of the rocks close to the fault at Ravenscar

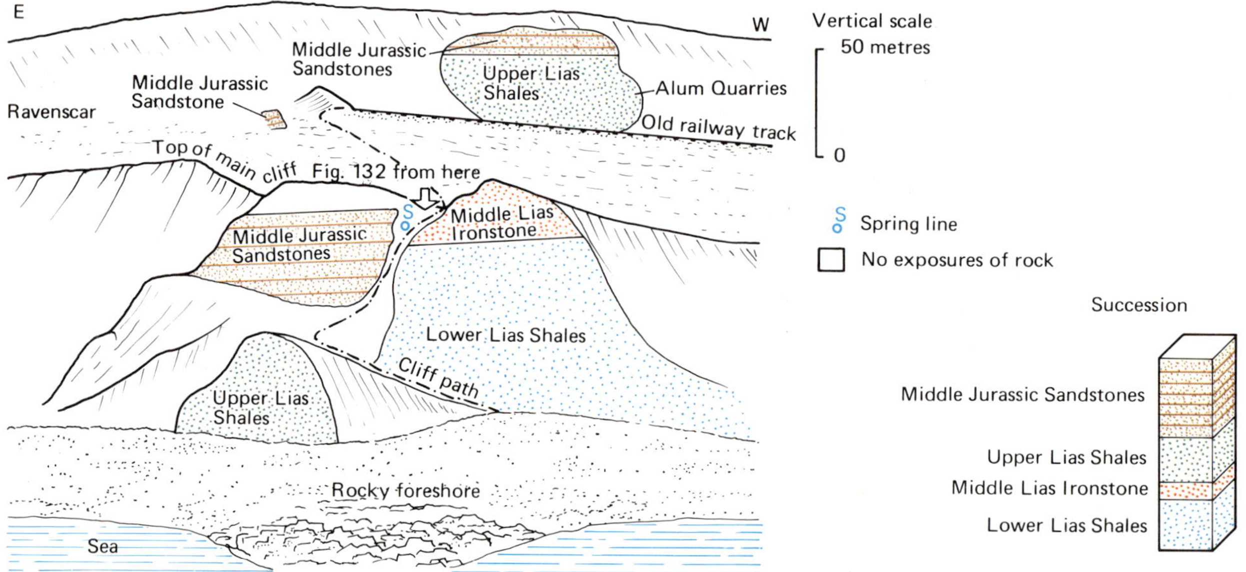

Fig. 132
The Peak fault is shown on the foreshore just before it splits into two branches

Fig. 133
Small faults in the cliffs near St Agnes, Cornwall

Fig. 134
A low-angle fault near Thurlestone, South Devon

- Make a trace of this field sketch.
- Mark on it the rough position of the fault in the cliff. (As in the Horton in Ribblesdale exercise (page 53), the spring line could be used as evidence that under it there is a boundary between two different rocks.)
- What is the throw of this fault?

The photo (Fig. 132) shows the rocky foreshore below the cliff at Ravenscar. The Peak fault, as it is called, splits into two branches here.

- Find these two branches on the photo.

The left-hand (west) branch of the Peak fault has a throw of about 100 metres and the right-hand (east) one a throw of about 30 metres. In each case, the downthrow side is to the east.

- Is the total throw of these branches about the same as the throw of the single fault in the cliff itself?

There are some other faults on the foreshore with much smaller throws than these.

- Can you spot any of them on the photo?

The faults in Fig. 133 also have very small throws.

- How can you tell this?

The Hunmanby fault and the Peak fault both have a near-vertical fault zone. The fault in Fig. 134 is different because it dips much more gently.

Also, these faults (including the one in Fig. 134) involve up-and-down movements in the rocks. But sometimes rocks can be 'torn' apart by mainly sideways movements.

The best-known modern example of a tear fault is the San Andreas fault in California. This is 1300 kilometres long and has a zone of broken rocks almost 2 kilometres wide.

Each sudden movement of this fault causes earthquakes, some of which have had very serious effects (page 21). The total movement caused by all these jerks is large. Every million years, the land on either side of the fault is separated by at least 4 kilometres, and sometimes up to 50 kilometres.

The rocks along a fault zone can keep on moving from time to time over many millions of years. But there always comes a time when they don't move again. Most of the faults in Britain are no longer active, though some still must be because minor earthquakes happen even here (page 21).

Further study 1

Holes in the ground

Deep holes and cracks in the earth like Gaping Ghyll (Fig. 1.1) are often found in limestone country. A hundred years ago farmers often used them to get rid of rubbish and the dead bodies of animals. They didn't know that there were underground rivers deep in the limestone. Their rubbish dropped into these in some places.

Fig. 1.1.
Gaping Ghyll, on the slopes of Ingleborough, North Yorkshire

A river could flow through the pile of rotting dead bodies, and then appear miles away as a spring. Villagers would drink the water thinking it was pure and clean.

This was one reason why the disease typhoid was common in some places in Europe in the 1800s. In France a keen geologist called Eduard Martel set out to prove that water at the bottom of some potholes flowed out in springs a long way off.

• How could he try to prove this?

In fact Martel poured green dye down the holes, and the springs miles away turned green a few hours later. As a result of his efforts a law was passed in France in 1902 forbidding people to dump rubbish down potholes or cracks.

Martel might have tried to climb down and follow the river underground. He actually did go down Gaping Ghyll and discovered the caves and tunnels shown on Fig. 1.2.

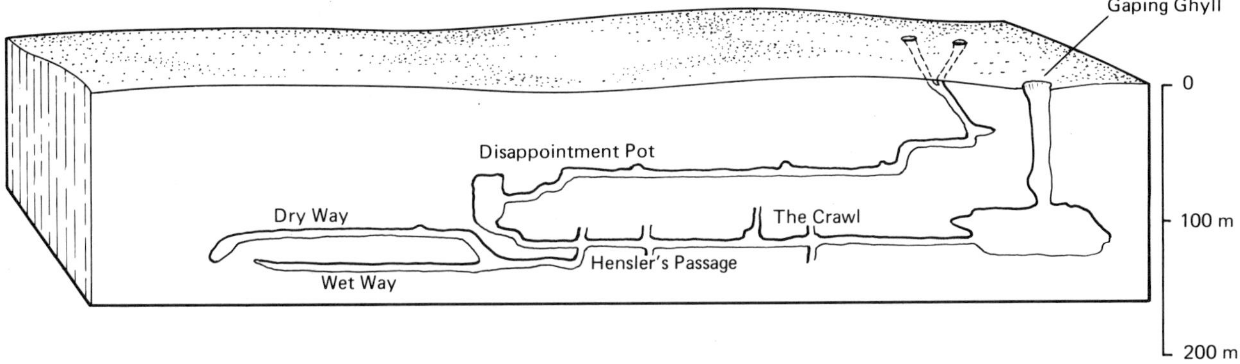

Fig. 1.2.
Some of the caves and passages down Gaping Ghyll

Limestone is famous for its caves, and some have been made safe for tourists to visit. Many others can only be reached after dangerous climbs, crawls and squeezes, and underwater swimming. The sport of potholing is quite popular in Britain now, with well trained and equipped potholers exploring underground.

Many wonderful limestone shapes are found in the caves and tunnels, like those in Fig. 1.3. These are made of limestone that has been deposited from water carrying a lot of dissolved limestone.

Fig. 1.3.
Inside a limestone cavern in Derbyshire

Fig. 1.4.
The work of water on limestone

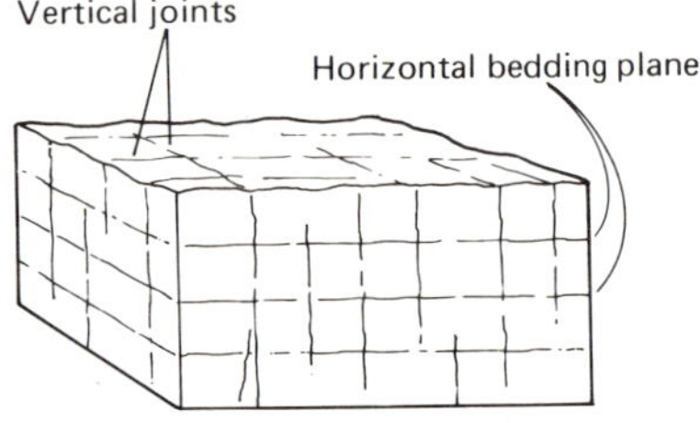

Fig. 1.4a
Limestone block before weathering

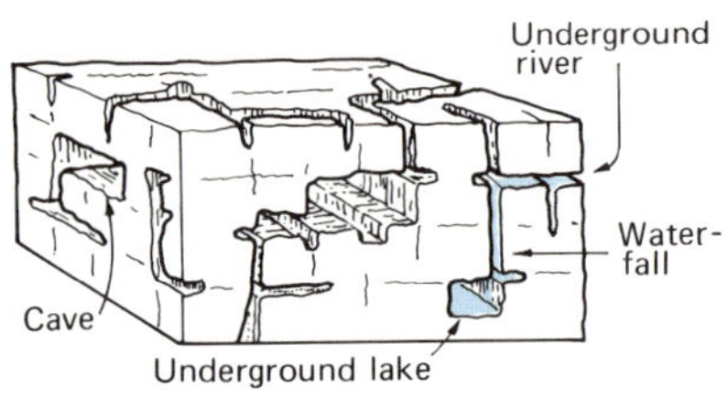

Fig. 1.4b
Limestone block after water has dissolved much of the rock

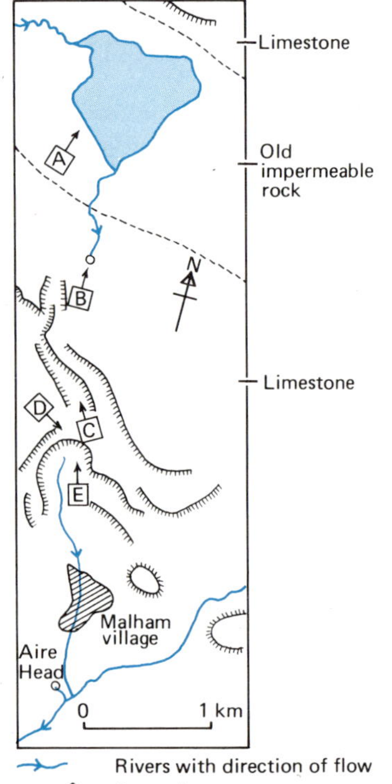

Fig. 1.5.
Limestone landscape near Malham

The water can dissolve the limestone rock in the first place because it is slightly acidic. Carbon dioxide picked up from the air by rain drops and the soil by soil water causes this acidity. This is a special form of weathering associated with limestone only. So you can't find potholes, and underground caves and tunnels in any other rocks.

Which features in the cave in Fig. 1.3 were probably deposited by water as it

(i) dripped from the roof?
(ii) splashed onto the floor?

Most limestones are criss-crossed with joints and bedding planes (page 3 and Fig. 1.4a). Rain water and streams seep into these cracks and make them bigger by dissolving the rock. Eventually the limestone is honey-combed by tunnels, gulleys and caves. Rivers disappear down potholes (or 'sinks') and whole systems of streams, waterfalls and lakes are made underground (Fig. 1.4b).

In Britain most of the underground limestone features were probably made during the Ice Age that finished about 10 000 years ago. Warm periods lasting many thousands of years were common during the Ice Age. There would be a huge amount of water about during those times as the thick ice sheets melted.

The caves in Britain are mostly found in Carboniferous Limestone that was deposited about 300 million years ago. So layers many hundreds of metres thick were probably dissolved away before the Ice Age even began.

One area famous for its limestone scenery is Malham in the North Yorkshire Pennines. The map (Fig. 1.5) shows some of the features near Malham. The photos (Figs. 1.6 to 1.10) were taken at the five places marked A,B,C,D and E on the map, but not in that order.

- What might be found if the stones in the foreground of Fig. 1.8 were dug away?
- Choose the right location (A,B,C,D or E) for each photo.

Fig. 1.6 shows a feature that is really typical of limestone scenery. Steep-sided valleys and gorges with rock outcrops along each side are often found in limestone country. And many of them, like the one in Fig. 1.6, have no river flowing along them now. The water that flows out from Malham Tarn disappears underground and reappears 2½ kilometres away at Aire Head.

Fig. 1.6.

Dry valley near Malham, with steep sides and many rock exposures

Fig. 1.7. Malham Cove, once a waterfall

Fig. 1.8. A disappearing stream

Fig. 1.9. Malham Tarn

Fig. 1.10. A limestone pavement. Gullies follow the lines of joints dissolved out by rainwater

Geologists agree that at some time water must have flowed across the land in places like Fig. 1.6 to make the valleys. But they don't agree exactly why or how this happened.

There are two main possible reasons suggested.

(i) During and after the Ice Age the ground must have been frozen solid to a great depth. When the climate got warmer the ground would only thaw out very slowly. Water from melting ice or rain would not be able to soak into the ground or flow down 'sinks' that would be blocked with ice. So the rivers would flow on the surface and cut valleys in the thawed-out top layers of the land.

(ii) At some times there has been much more water to drain off the land than today. This could be because of melting ice, or a different climate with heavier rainfall. Because so much water could not drain away fast enough the caves and tunnels were flooded. The rivers had to flow on the surface as well as underground.

In 1969 after a very heavy storm a great torrent of water flowed down Cheddar Gorge in the Mendip Hills. Usually the gorge is dry, and the flood ripped up the road that had been built along the bottom of the valley.

- Which of the two explanations for dry valleys does this event seem to support?

Though it is agreed that water eroded the dry valleys, each one presents another problem. The valley could have been eroded in the normal way by a river on the surface. But it could have been eroded by a river underground, and then the roof of the tunnel could have fallen in.

- What kind of evidence would you look for to prove that a valley was once a tunnel? Think about (i) the shape of the valley; and (ii) evidence from the valley floor.

It is difficult to be certain about the origin of any dry valley in limestone.

- What do you think about the one at Malham?

Further study 2

The birth and death of a volcano

No-one knows exactly when Surtsey was born. A volcano that erupts on the sea floor isn't noticed until its cone gets close to the surface of the sea.

The first sign—a strange smell of sulphur— was detected in the early morning on November 14th, 1963 by the crew of a fishing boat off the south coast of Iceland (Fig. 2.1). Soon afterwards, the boat started to roll, the sea went a brownish green colour and smoke billowed up into the air several kilometres high. Luckily the boat wasn't sailing over the very top of this new volcano.

An island rose up from the sea during the night (Fig. 2.2).

Submarine eruptions are much more explosive than eruptions on land. This is because cold sea water can get into the vent and so mix with the hot magma. The explosion hurls a dense mass of fragments out of the vent. The leading fragments may trail smaller fragments behind them, and from a distance these may look like rockets.

The darker colours in any volcanic cloud are caused by solid fragments (tephra) varying in size from fine-grained ash up to large blocks. The coarse tephra gets deposited on the cone close to the vent, but some of the finest tephra may be blown by the wind and deposited much further away.

Read the eyewitness account (Fig. 2.3) of a daring landing made by rubber dinghy one day in February 1964 while the volcano was still very active. It shows that tephra contains molten material as well as solid material.

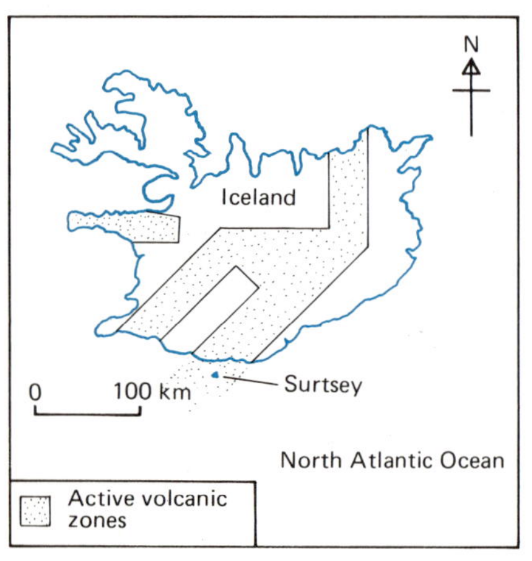

Fig. 2.1.
Active volcano zones in Iceland

Fig. 2.2.
The new volcano after one week of activity

'At first after we stepped ashore, Surtsey was quiet. We kept our fingers crossed in the hope that it would stay like this. But we had not been there many minutes when the vent began to fire warning shots. Before we really could make out what was happening we saw water spouts in the sea off the beach where we were standing. They came from bombs of molten lava crashing down, and these soon began to fall all around us.

Under such circumstances, there is only one thing for you to do. Instead of taking to your heels, you have to stand still and stare up in the air, trying not to dodge the bombs until the very moment they seem about to land on your head. This is really not quite as hard as it might seem.

When the biggest bombs, almost a metre in diameter, crashed on the wet sand, there was a bang that sounded rather uncanny at close range. On landing they cupped out a hole in the sand that soon filled with water, which boiled against the red-hot lava chunk.'

From: S. Thorarinsson, *Surtsey*

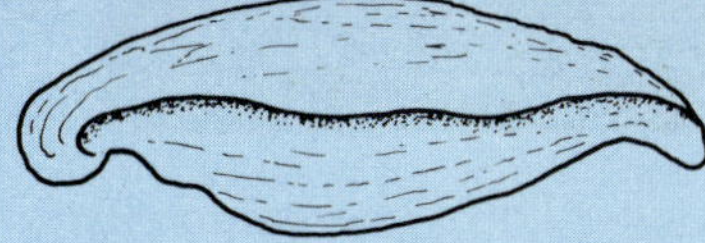

One kind of volcanic bomb

Fig. 2.3.

You will guess that this visit was soon ended and that no more attempts were made to go ashore on Surtsey until things had quietened down a bit!

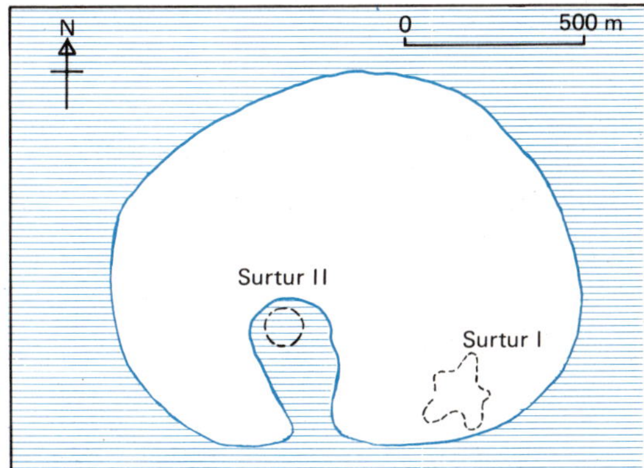

Fig. 2.4.
Surtsey on 17th February 1964

The shape of the island around the time of this landing is given in Fig. 2.4. But then, a new vent (Surtur II) was opening up quite close to the old one (Surtur I).

- How is the position of Surtur I on the island linked to the wind having mostly come from the south during the first two months of the volcano's life?

All this time the sea was attacking the newly formed rock. Tephra is easily eroded. Once a volcano like this stops erupting, the island gets quickly smaller and may finally disappear altogether.

A good example was a little volcanic island that appeared off the coast of Surtsey on May 28th, 1965. Fig. 2.5 is a diary of the events that followed. Notice what an important part the weather played in this story.

But Surtsey was bigger and didn't die in this way. During the early part of 1964, it became big enough for Surtur II to be protected against the sea. And on April 4th the red-hot lava started to spill over the top of the vent towards the sea (Fig. 2.6).

Fig. 2.5.
Some stages in the life of a small volcanic island off Surtsey

17 metres
Sea
June 8th
Storms
Island 'disappears' but volcano carries on erupting under the sea

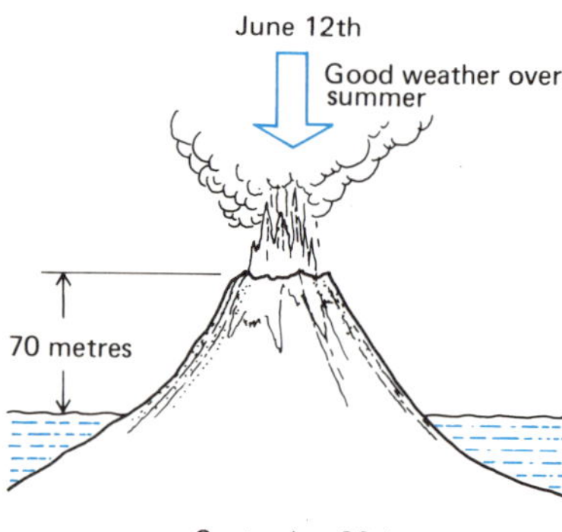

Island 'disappears' for the last time
October

Fig. 2.6. The lava flow at night. The white cloud on the left is formed over the place where red-hot lava meets the sea

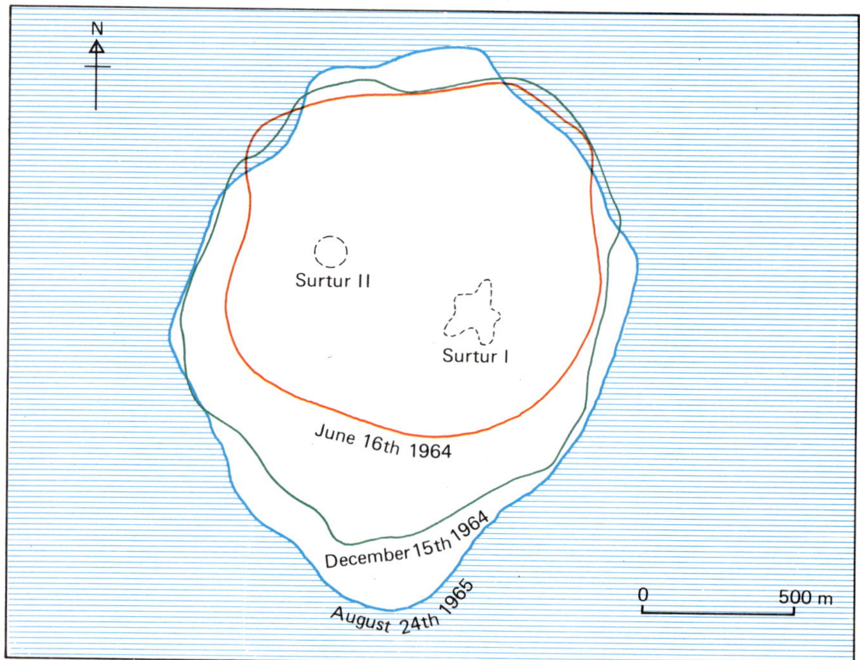

Fig. 2.7.
The growth of Surtsey during 1964 and 1965

Fig. 2.8.
Surtsey in 1965

This lava flow made certain that the island would survive for many years because lava rock is much less easy to erode than tephra rock. In fact, Surtsey got quite a lot bigger during 1964 and 1965 (Fig. 2.7).

- Make a trace of Fig. 2.7 and put it over Fig. 2.4 so that Surtur I and Surtur II are matched up.
- In which direction did the lava move from Surtur II?
- Did more lava pour out of the vent in 1964 or in 1965?

Surtsey now looked like the photo (Fig. 2.8). There was a constant battle between the advancing lava and the wind-whipped waves of the sea.

- What effect would the sea have on the surface of the hot lava?
- What can you see in Fig. 2.8 which shows that the hot lava makes the sea boil?

Some lava took the line of least resistance and turned to flow along the beach rather than straight out to sea.

The volcanic activity on Surtsey stopped in 1967. How long the island lasts depends partly on whether there are any more eruptions in the future. The sea will erode the tephra rock quickly and the lava rock more slowly.

Perhaps in a thousand years time, Surtsey will look like one of the many small stacks that you can see in the distance in Fig. 2.8—the last remnant of a dead volcano.

Further study 3

Making a match

Imagine that you are interested in developing a new coalfield and you want to know where to find the thickest seams of good quality coal. The first thing to do is to drill boreholes in a few places.

In each borehole, you will come across several seams of coal at different depths. These then have to be matched up (correlated) across all the boreholes. In this way, you can get a kind of 3-D picture of the rock patterns beneath the surface and you can decide where best to start mining.

The only fossils in the coal seams themselves are the microscopic pollen grains. These can be used in correlation. But there is another way using the invertebrate fossils in some of the other rocks found along with the coal in the Coal Measures.

Useful rocks for correlation are ones that occur in a lot of places and can easily be recognized by their fossils. A particular set of fossils helps to 'mark out' such a marker band. The coal seams just below or just above a marker band can then be matched up in different boreholes.

In the Coal Measures, two kinds of marker band are the marine bands and the mussel bands. The marine bands contain the fossils of animals that once lived in the sea while the mussel bands contain the shells of lamellibranchs that lived in water which wasn't as salty as sea water.

The Selby area in North Yorkshire is being developed as a new coalfield. Fig. 3.1 gives some information from three of the boreholes drilled near Selby—at Kelfield, Barlow and Whitemoor.

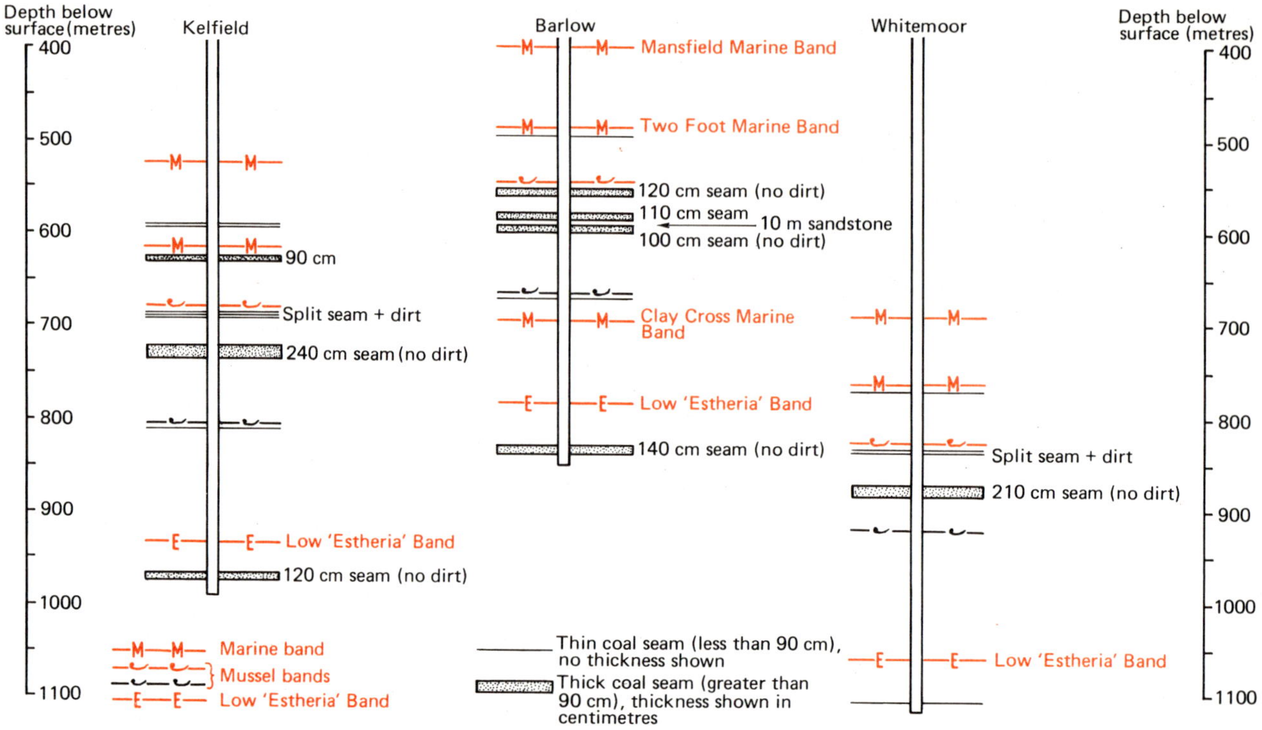

Depth below surface (metres)
400
500
600
700
800
900
1000
1100
Kelfield
90 cm
Split seam + dirt
240 cm seam (no dirt)
Low 'Estheria' Band
120 cm seam (no dirt)
Marine band
Mussel bands
Low 'Estheria' Band
Barlow
Mansfield Marine Band
Two Foot Marine Band
120 cm seam (no dirt)
110 cm seam
10 m sandstone
100 cm seam (no dirt)
Clay Cross Marine Band
Low 'Estheria' Band
140 cm seam (no dirt)
Thin coal seam (less than 90 cm), no thickness shown
Thick coal seam (greater than 90 cm), thickness shown in centimetres
Whitemoor
Split seam + dirt
210 cm seam (no dirt)
Low 'Estheria' Band
Depth below surface (metres)

Fig. 3.1. Some geological information from three boreholes near Selby

Two Foot Marine Band
Clay Cross Marine Band
Mansfield Marine Band.

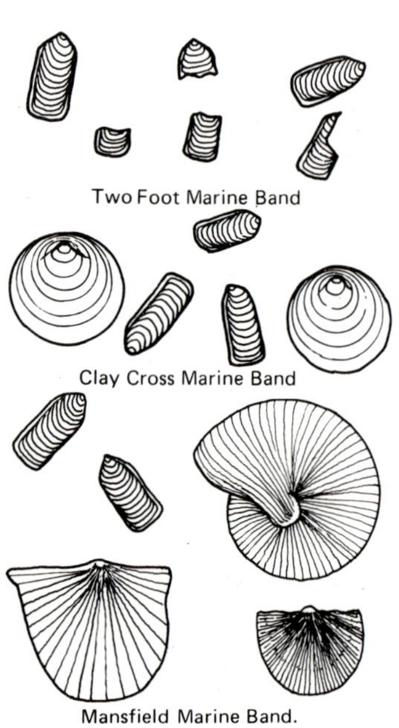

Fig. 3.2. Some fossils from the marine bands

There are three marine bands in the Barlow borehole, and some of the fossils that can be found in each of these bands are drawn in Fig. 3.2.

The marine band at 525 metres in the Kelfield borehole and the one at 685 metres in the Whitemoor borehole both have the same kinds of fossils (Fig. 3.3).

- Is the upper marine band at Kelfield and Whitemoor (i) the Mansfield Marine Band, (ii) the Two Foot Marine Band, or (iii) the Clay Cross Marine Band?
- What is the lower marine band at both Kelfield and Whitemoor?
- Which marine band at Barlow is missing in both the other boreholes?
- Is the split seam at 585 metres in the Kelfield borehole also found at Barlow or Whitemoor?
- Can you match up the seam at 480 metres in the Barlow borehole with a seam in the other two boreholes? On a copy of Fig. 3.1, draw lines joining up these matched seams.

Besides the marine bands, three other marker bands are passed through in all the boreholes. These can help you to match up some more coal seams.

- Are these three other marker bands marine bands, mussel bands or *Estheria* (a fossil crustacean) bands? Draw lines joining up the same bands in each borehole.
- Now draw lines joining up the seams in each borehole that can be matched up.

One of the seams at Barlow isn't passed through at Whitemoor.

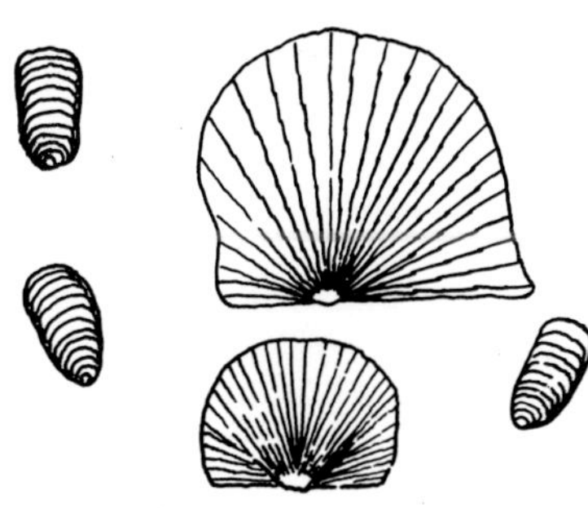

Fig. 3.3.

- At what depth is this seam passed through in the Barlow borehole?

The thickest seam (the so-called Barnsley seam) in the Kelfield borehole is the 240-centimetre seam at a depth of 730 metres.

- Is the Barnsley seam also the thickest seam in the Whitemoor borehole?
- What is different about the Barnsley 'seam' in the Barlow borehole?

It is the Barnsley seam that the National Coal Board decided to mine in the new coalfield, though only where it can be found as a thick single seam.

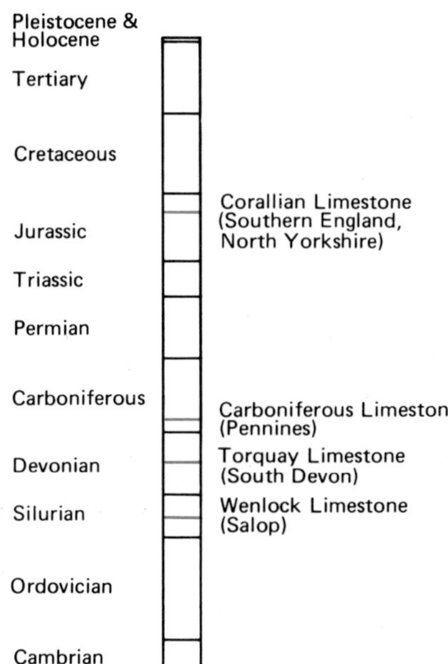

Fig. 4.1.
The coloured line marks a limestone made partly from ancient coral reefs

Further study 4

Past environments

Sedimentary rocks and their fossils can be used to tell us a lot about the conditions over the British area through earth's history. Often this is done by comparing these ancient materials with modern sediments and living things.

Such work shows that there have been some very dramatic changes in the pattern of land and sea, and in the climate. So coral seas have given way to tropical swamps, and these in turn to high mountains and hot deserts. One of the latest of these changes brought glaciers and ice sheets to Britain during the Pleistocene.

Fig. 4.1 shows the times when limestones made partly from ancient coral reefs were formed over Britain. There are no coral reefs around our coasts nowadays and so conditions must have been very different then.

Modern coral reefs only thrive in certain parts of the world's oceans (Fig. 4.2).

- Outside which latitudes north and south of the equator are there no coral reefs at all?

The idea of the British area once being a warm tropical sea with coral reefs doesn't sound very likely. There are at least three ways of trying to explain the coral limestones.

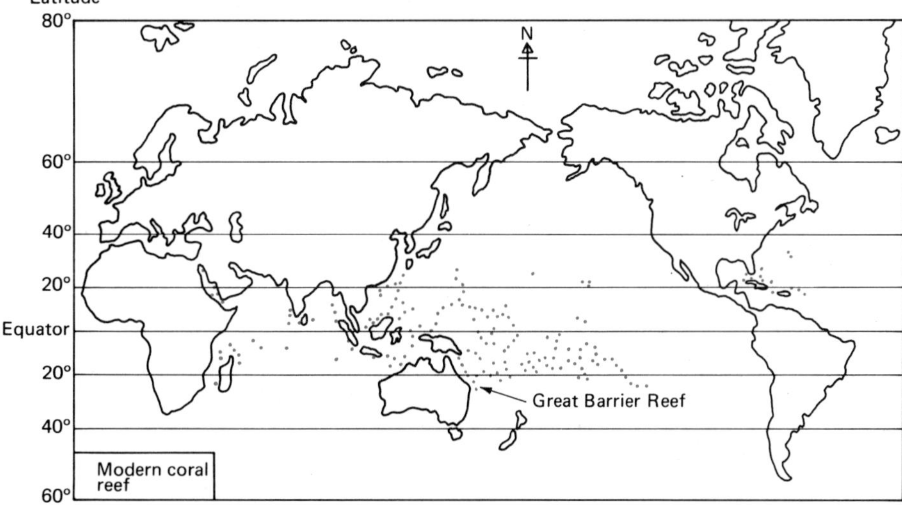

Fig. 4.2.
Distribution of the world's coral reefs

1. The ancient corals were able to live in cooler seas than the modern corals.
2. The ancient seas were generally warmer than the modern ones, and so corals could live within a wider range of latitudes.
3. Britain has sometimes been much closer to the equator.

There could be something in all of these explanations but it is the third one that geologists like best.

Not all limestones of Carboniferous age are coral limestones. But they are mostly very rich in other fossils like crinoids and brachiopods, and this fact also suggests a warm sea rather than a cool one.

Conditions must have changed later in the Carboniferous period because these limestones become less common. Instead the important rocks are sandstones and shales along with rootlet beds and coal seams. The rocks in Swaledale (Fig. 32) are a mixture of limestones and these other rocks.

Many of the fossils in these rocks are plant remains—roots, leaves, stems and pollen grains. Obviously the British area was land for most of that time, though the presence of ancient river channels (called washouts, Fig. 4.3) in the rocks show that the land was swampy and not very dry.

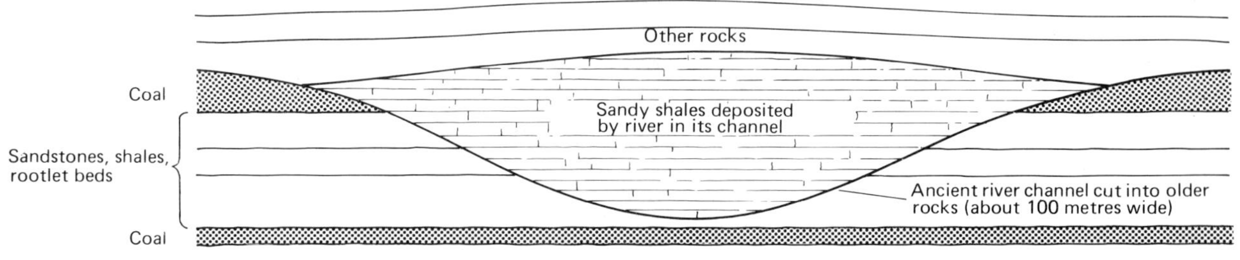

Fig. 4.3.
A section through a washout

Coal seams are formed from huge amounts of partly rotted vegetation that have been quickly buried under piles of sediment. The area needs to be one where there is plenty of sun, high temperatures and heavy rainfall so that the plants and trees grow very quickly. And it also needs to be one where sediments are being deposited.

The Ganges delta in India is a modern example of such an area. It is made from alternate layers of peat and sand that eventually could form coal seams and sandstones.

- Why are rocks formed by a delta likely to contain many washouts?

Fig. 4.4 is an artist's impression of what the Upper Carboniferous swamps may have looked like. Though these swamps replaced the coral seas, the climate stayed very tropical all through the Carboniferous period.

Now let us look at the younger rocks of the Permian and early Triassic periods. Several of these are red sandstones with large-scale cross-bedding (see Fig. 19) and very well-rounded grains of sand. The cross-bedding could be formed under water or in a desert sand dune. But the very well-rounded grains are more likely to have been formed as they were 'blasted' against rocks by strong desert winds.

So it looks as if these red sandstones were formed in a hot sandy desert like the one in Fig. 4.5. Parts of the Sahara Desert are a modern example.

- Does the lack of fossils in the red sandstones also fit with this idea?

Fig. 4.4. An Upper Carboniferous swamp

Fig. 4.5. Part of the Permian desert

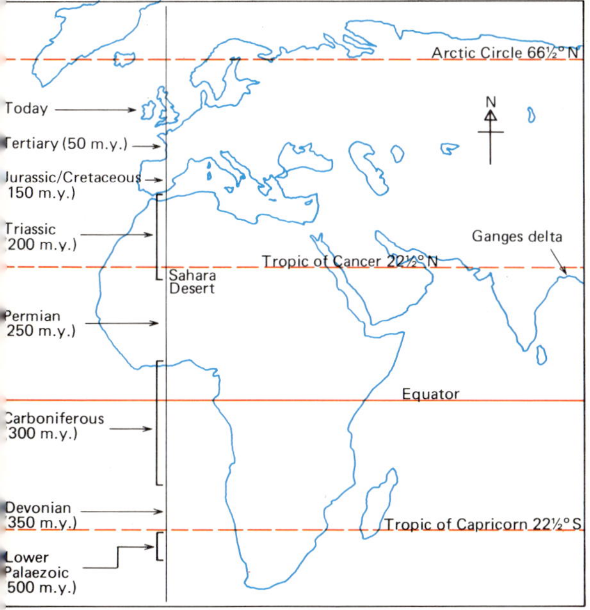

Fig. 4.6.
The changing latitudes of Britain over geological time

This raises the same problem as for the coral seas and tropical swamps. Again, the explanation could be that Britain was once in a different latitude—perhaps the one that the Sahara Desert is in now. Geologists think that Britain has been in different latitudes at different times (Fig. 4.6). It may have started off south of the equator in Lower Palaeozoic times and moved quite quickly across the equator during the Carboniferous period. Since then it has carried on moving northwards. This movement is, of course, connected with the rafting of whole continents across the earth's surface on the backs of very large mobile 'plates' (page 23).

So, since the Carboniferous, the British climate has slowly cooled down. During the Pleistocene it was very cold indeed here for some of the time, but this was because the whole earth became much colder and not because of the drift to the north.

The evidence for Britain's 'voyage' across the earth's surface comes partly from a study of the magnetism of rocks. But the past environments that have been worked out using sedimentary rocks and fossils do seem to match the latitudes in Fig. 4.6 very well.

Further study 5

Rock cycles in the past

The rock cycle has been going on in the past just as it still goes on today. And an unconformity is one important clue in the rocks that this must be so.

During the long time-gap represented by an unconformity, the older set of rocks was tilted or folded and then uplifted well above sea level. These events are shown in Figs. 56 and 60 for a place where the folding is intense and the uplift produces high mountains. Lastly, the new land is completely eroded so that rivers or seas can move across it and deposit the younger set of rocks.

You can use an unconformity to work out a very rough date for the time when the new land was being formed. Look at the photo of the unconformity near Oban in western Scotland (Fig. 5.1).

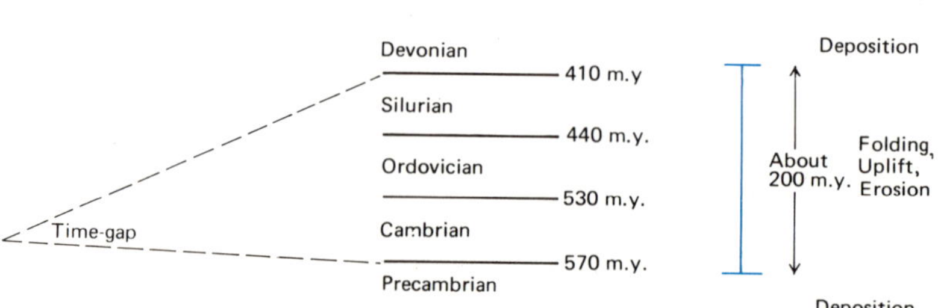

Fig. 5.1.
The unconformity near Oban in western Scotland

The conglomerate is of Lower Devonian age and the highly folded rocks below were deposited in late Precambrian or early Cambrian times. This means that the time-gap is about 200 million years, and mountain-building (uplift) was only going on for just part of that time.

In fact, there is a different way of getting a date for the mountain-building at Oban. The rocks under the unconformity are metamorphic, and radiometric ages have been found for some of them. Each age shows when the process of changing the rocks (metamorphism) is over. This is when the temperature drops and some of the pressure is released.

Radiometric ages for these rocks in Scotland are mostly between 490 and 430 million years. Suppose that the rocks were metamorphosed and uplifted at around the same time.

- In which geological period did the mountain-building at Oban probably take place?

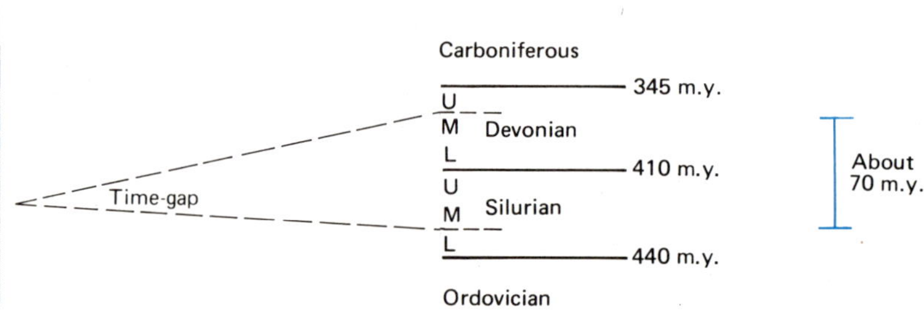

Fig. 5.2.
The unconformity at Siccar Point in south-east Scotland

Now look at the unconformity in Fig. 5.2. This is exposed at Siccar Point on the east coast of Scotland close to the border with England. Here Upper Devonian sandstone lies on steeply dipping Lower Silurian rocks. So the time-gap is about 70 million years.

The date for the mountain-building at Siccar Point can't be the same as for Oban.

The unconformity at Horton in Ribblesdale (page 54) is very like the one shown in Fig. 5.3 near Llangollen in Clwyd, north Wales. Lower Carboniferous limestone lies on folded Ordovician and Silurian rocks.

- Work out the smallest time-gap for this unconformity.
- In which geological period did mountain-building probably take place at Llangollen?
- Could this event at Llangollen have happened at about the same time as the mountain-building at Siccar Point?

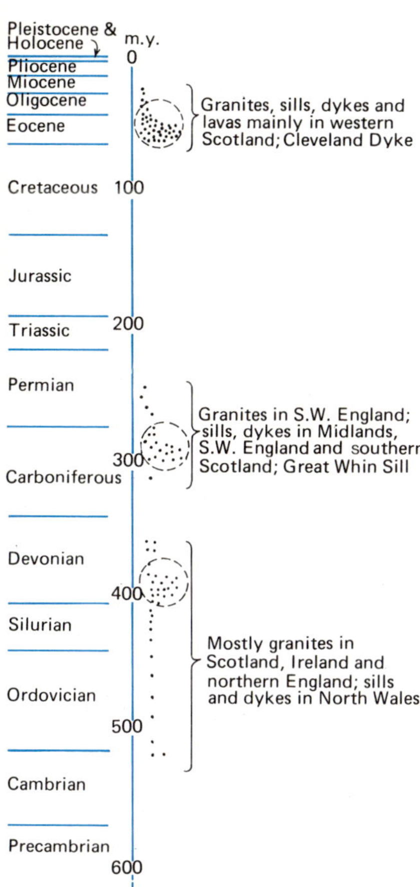

Fig. 5.4.
Each dot represents the radiometric age for an igneous rock in Britain or Ireland

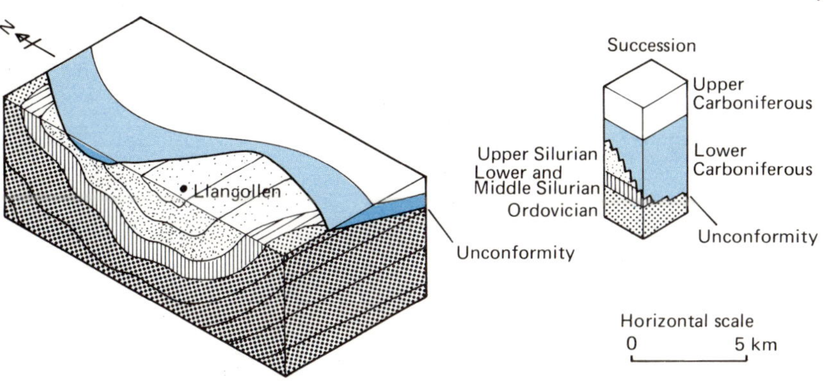

Fig. 5.3. The Carboniferous–Silurian unconformity at Llangollen in Clwyd, Wales

Granite plutons and dolerite sills and dykes are often found in folded rocks. For these, the radiometric age is the time when the molten material cooled to form crystals of solid rock.

Fig. 5.4 shows the radiometric ages for a large number of igneous rocks in Britain and Ireland. Notice that they seem to group around three ranges of time.

- Which are these?

Look for any links between the dates of mountain-building at Oban, Siccar Point and Llangollen and the ages of these igneous rocks. There doesn't seem to have been very much igneous activity at the time when one of these unconformities was being formed.

- Which one is this?

But there was a lot at the time of the other unconformities.

You will see from Fig. 5.4 that there was another burst of igneous activity just under 300 million years ago. Is this also linked to more folding and mountain-building?

Figs. 5.5 and 5.6 show an unconformity between Triassic and Carboniferous rocks near Wells in Somerset.

- Work out the time-gap for the unconformity in each figure.
- In which geological periods could mountain-building have taken place?

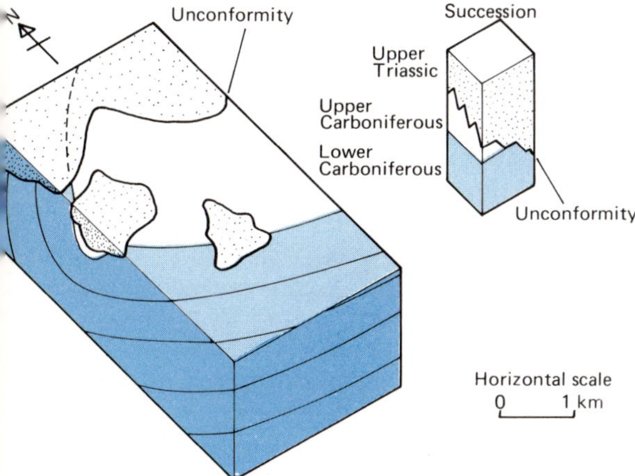

Fig. 5.5.
The Triassic–Carboniferous unconformity near Wells in Somerset

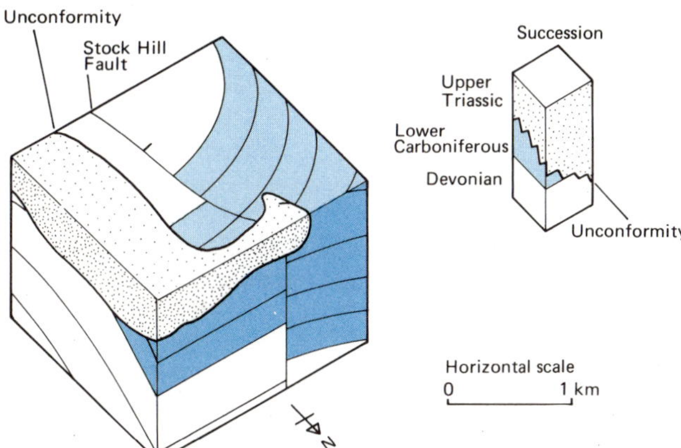

Fig. 5.6.
A fault under the Triassic–Carboniferous unconformity near Wells in Somerset

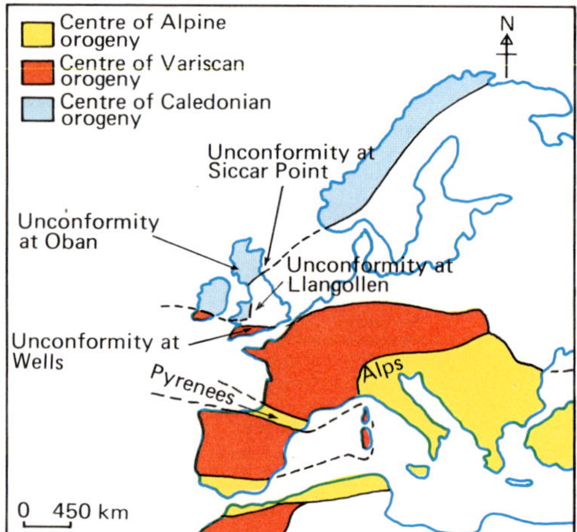

Fig. 5.7.
The centres in Europe for the Caledonian, Variscan and Alpine orogenies

Mountain-building movement (orogeny)	Time range	Time of main uplift in Britain
Alpine	Late Cretaceous to Pliocene	About 20 m.y. ago
Variscan (Armorican or Hercynian)	Late Devonian to early Permian	About 300 m.y. ago
Caledonian	Late Cambrian to mid-Devonian	About 400 m.y. ago (England, Wales, southern Scotland); about 470 m.y. or 530 m.y. ago (Highlands of Scotland)

Fig. 5.8. Three important mountain-building movements.

- How well does this event match the burst of igneous activity? There is also a fault in Fig. 5.6.
- How can you tell from this that the fault was formed later than the folding?

Even so, the fault must have been formed during the mountain-building and before the erosion of the older rocks.

- What proves this in Fig. 5.6?

There have been three main mountain-building movements (orogenies) since the start of the Cambrian (Fig. 5.8). Each one lasted many millions of years but reached a peak for just a few million years.

These orogenies had their greatest effects in different places (Fig. 5.7). Only during the Caledonian and Variscan orogenies was a part of Britain at the centre of things. The mountains thrown up by these may once have been as high as the Alps are today.

Rocks at the centre of the 'storm' were metamorphosed and intensely folded and faulted. This is what happened to most of the rocks in Scotland during the Caledonian orogeny.

If the rocks were quite a long way from the centre, only ripples from the main movements were felt. The rocks weren't metamorphosed, but they were often tilted and sometimes folded and faulted. This is what happened in Britain during the Alpine orogeny, which had its centre much further south.

The anticline at Albury (page 56), the Wealden anticline and the London syncline were all caused by ripples from the Alpine orogeny. On the south coast of England some of the rocks were quite strongly folded (Fig. 5.9).

Fig. 5.9.
An aerial view of the western end of the Isle of Wight

- What does Fig. 5.9 tell you about the date of the folding of these rocks?

Though the British mountains are made of very old rocks, their present height has a lot to do with the uplift caused by the Alpine orogeny. The mountains of the Caledonian orogeny have long since been eroded away to their 'stumps'.

The dates for some of the igneous activity in Britain (Fig. 5.4) match well with the peaks for the Caledonian and Variscan orogenies. This is because Britain was close to the centre of things at both times.

- Look back at Fig. 57, and decide for each of these two orogenies which two continents were in collision over what are now parts of Britain.

The match isn't quite so good for the Alpine orogeny. Fig. 5.7 shows that the centre of this orogeny was a very long way from the igneous activity in western Scotland. It is obvious that the igneous activity wasn't a direct result of the squeezing of the rocks in southern Europe. So this must have been caused by something else.

- Use Fig. 57 to find out what important event was taking place around this time over western Scotland.

Further study 6

Geology and motorways

Building a motorway like the one in Fig. 6.1 is a complex business. Knowing about the soils and rocks beneath the surface of the land around the chosen route is vital if mistakes aren't to be made.

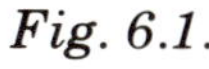

Fig. 6.1.

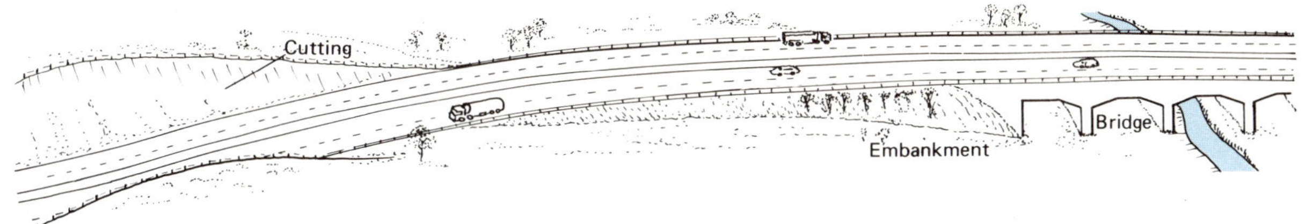

In the cutting the slopes of the banks must not slip into the road when they get very wet after a lot of rain. If the cutting goes through rocks, the road engineer has to be sure that weathered bits of them won't break off and fall into the road.

On the embankment, the problems are different. Here the material used to build it must be firm enough, even in the wettest weather, not to move under the weight of all the traffic on the road.

Foundations are also important for the bridge over the river. The floodplain of the river may be partly made from loose materials like sands or silts. These are easily pressed down by heavy weights and so foundations built on them would soon begin to sink (along with the bridge!).

The problem can be solved if there is a firm clay not far below the surface. The loose material can then be dug out so that the foundations are built on the clay, or piles can be driven through it to the clay.

But deciding on the best route for a motorway doesn't only depend on the geology. Motorways affect people in various ways. In the country a farmer might lose some of his land and have to use a motorway bridge or an underpass to get from one part of his farm to another. In towns, houses have to be knocked down to make way for the road, and people living right next to the road have to get used to a lot more noise and exhaust fumes.

It isn't easy to decide about a motorway route, especially in a town. Often the best route for the road engineer isn't the best for the people living in the area. The arguments on both sides have to be weighed up very carefully.

The south Wales town of Port Talbot (Fig. 6.2) used to have serious traffic problems, and a motorway section for the A48 road was planned. (This is now part of the M4 from London to south

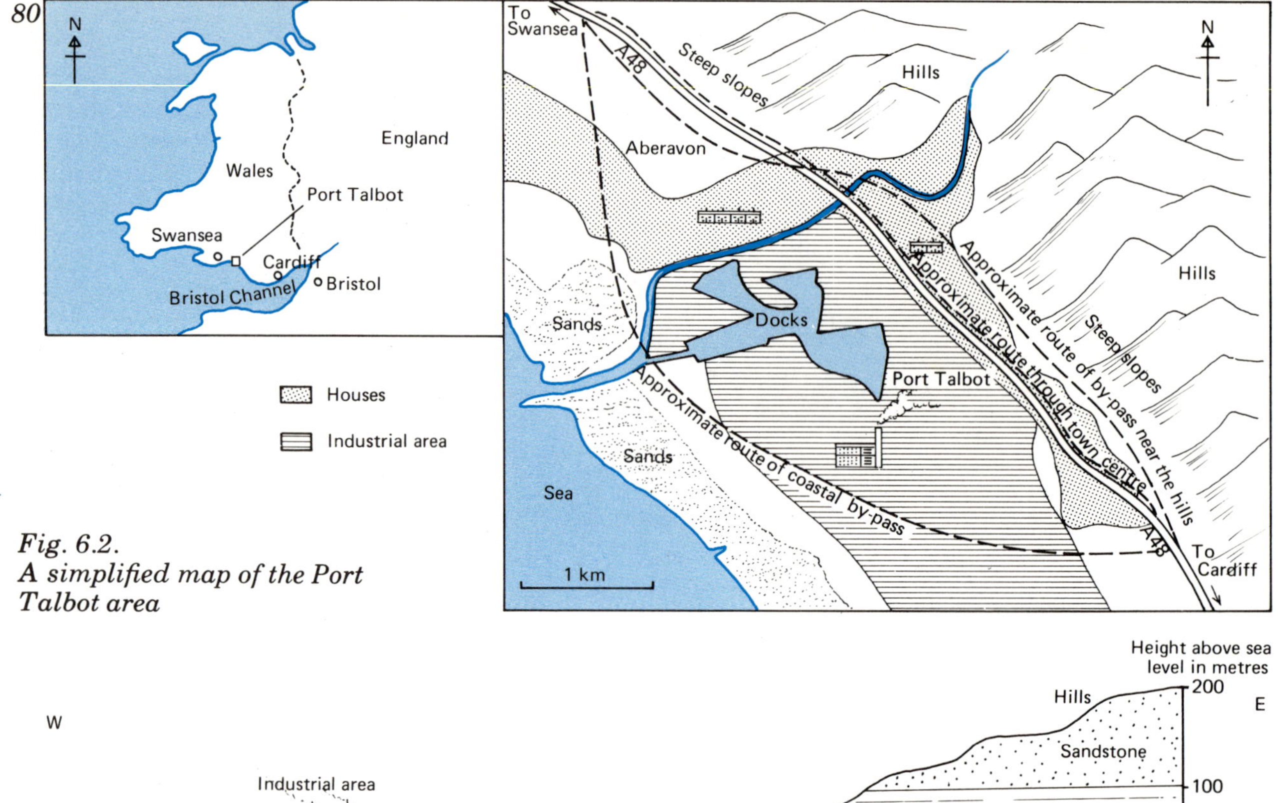

Fig. 6.2.
A simplified map of the Port Talbot area

Fig. 6.3.
A section through the Port Talbot area

Wales.) There were three possible routes—a new road roughly along the line of the old road through the centre of the town, a by-pass along the coast, and a by-pass near the hills to the north-east of the town.

The diagram (Fig. 6.3) shows the geology of Port Talbot. The road through the town would have a firm clay foundation whereas the coastal by-pass would have to be built on loose materials like sand, silt and peat.

- Which one of these two would make the better foundation?
- What would be the problems of building a motorway along the old A48?
- What would be the problems of building a motorway along the coast?

There are landslips on the lower slopes of the hills behind Port Talbot. Any motorway here would have to be built on the shales and blocks of sandstone which have slipped down the hillside.

- What might happen to the landslipped material once there was a road with heavy traffic on it?

You will now realize that one of the three routes is best from an engineering point of view, and another one is best because it would have the least effect on houses and industries.

- Along which route do you think the motorway should have been built?

You could find out which route was actually chosen by looking at a map of the Port Talbot area.